The Great Balancing Act

THE GREAT BALANCING ACT

An Insider's Guide to the Human Vestibular System

JEFFREY D. SHARON

Columbia University Press
New York

Columbia University Press
Publishers Since 1893
New York Chichester, West Sussex

Library of Congress Cataloging-in-Publication Data

Names: Sharon, Jeffrey D. author
Title: The great balancing act: an insider's guide to the human vestibular
system / Jeffrey D. Sharon.
Other titles: Insider's guide to the human vestibular system
Description: New York: Columbia University Press, [2025] | Includes
bibliographical references and index.
Identifiers: LCCN 2025008626 (print) | LCCN 2025008627 (ebook) |
ISBN 9780231218627 hardback | ISBN 9780231218634 trade paperback |
ISBN 9780231562379 ebook
Subjects: LCSH: Vestibular apparatus | Vestibular apparatus—Diseases
| Equilibrium (Physiology)
Classification: LCC QP471 .S53 2025 (print) | LCC QP471 (ebook)

Cover design: Henry Sene Yee
Cover illustration: Jeffrey D. Sharon

GPSR Authorized Representative: Easy Access System Europe, Mustamäe tee 50,
10621 Tallinn, Estonia, gpsr.requests@easproject.com

Contents

The Great Balancing Act

Introduction

A Sense of Wonder

What in the world is a vestibular system? Why is it important? What happens to someone when their vestibular system stops working? In this book, we are going on a journey together to develop an understanding of the human vestibular system. We'll cover philosophy, physiology, and pathology. We'll try to understand the mechanics of the vestibular system, at the level of individual cells, neuronal circuitry, and behavior. We'll need to go deep into the skull to understand anatomy and function. We'll cover history and discuss groundbreaking ideas with brilliant scientists. And we'll meet several patients who will help us understand the personal experience of vestibular malfunction.

To get started, let's try an experiment. Stand up. Two legs on the ground, eyes open. How does your balance feel? Most people would say pretty good. Now, let's make things more difficult. Stand on a pillow. If you've got a thick, foamy pillow, you'll notice that balancing is harder now. Still too easy, you say? Well then go ahead and close your eyes (and needless to say, be careful). Balancing on a thick pillow, with eyes closed, is challenging for many people. If you really want to challenge yourself, balance on one leg, on a pillow, with your eyes closed. You have better balance than 95 percent of your friends if you can do that!

Why does balance become more difficult as we close our eyes? Balance relies on information from several sensory systems, which are then integrated within our brains. As you would surmise, vision is crucial for several reasons. First, it allows us to form a mental map of our surrounding area. Even while reading, you still have a general sense of the layout of the room that you are in. Second, vision provides constant feedback, so if you start veering or listing to one side you'll get the information you need to correct your balance. Third, as we walk around, we take advantage of our mental encyclopedia of surfaces and their material characteristics. For example, if you spot a patch of ice on the sidewalk, you alter your walking strategy to account for the much lower friction on the icy surface, in order to avoid sliding and slipping. So it's easy to see (pun intended!) why vision is important for balance.

So what about the pillow? Why does standing on a pillow make balancing harder? Intuitively, we understand that we need our sense of touch—especially on the bottom of our feet—to help with balance. The ability to sense things with our skin, muscles, or joints; including touch, temperature, pain, or vibration is called "somatosensation." Another specific form of somatosensation is called proprioception. Proprioception is our ability to know where parts of our bodies are in space. Meaning that if you consider a human to be an action figure, you could picture that our bodies can be moved into thousands of different poses. And, remarkably, even with our eyes closed, we know, at any moment in time, our own configuration. For example, we know if our left arm is outstretched, or is curled inward. We can sense that because of stretch receptors located in our muscles and our tendons. Since movement is based on muscle shortening, pulling on parts of our skeleton, which then rotate at our joints, by measuring the stretch of our muscles, our brains can deduce the position of each part of our body. Interestingly, our muscles can only contract—or get shorter. In order for them to go back to a previous position (i.e., to lengthen them), we need an opposing muscle to also contract, while the first muscle is inhibited. That is why our muscles come in opposing pairs, like biceps and triceps. Getting back to our experiment, we utilize the sensation of the floor beneath our feet (touch), and the position of different parts of our legs, like our feet, ankles, calves, and thighs (proprioception) to help balance. With certain diseases, one could lose somatosensation, which would make someone feel like they were constantly walking around on a mushy surface, like say an inflatable castle at a county fair. Which may be a bad example, because obviously there's nothing fun about losing such an integral sensation. One such disease is diabetes, which can cause "neuropathy" or nerve damage, resulting in foot numbness.

Now that we've seen how vision and sensation contribute to our sense of balance, it's time to introduce the vestibular system. The vestibular system is a part of the inner ear that senses gravity and head movements. The inner ear is a fascinating organ, housed deep within the skull, which contains that apparatus that "senses" hearing and balance. In future chapters, we'll explore the reason why hearing and balance are so closely entwined. We'll also review how exactly the inner ear is able to perform these extraordinary feats. For now, let's go back to our example. For those of you who were able to perform that hardest feat—standing on the pillow, on one foot, with your eyes closed—how were you able to do it? Your sense of vision was removed, and your sense of sensation was quite diminished. However, you were still able to balance. That is because of the vestibular system. By sensing gravity in order to know which way is up, and by sensing small changes in head tilt as you sway to one side, you are able to make small adjustments in muscle tone and position to compensate, keeping your center of gravity in a stable position, and preventing you from falling.

Center of gravity refers to a single point where the average mass of a person (or object) lies. So, the exact position for a complex object like a human being depends on the weight of all the various components of a person. So, here are two fun ways to visualize the center of gravity. The first is to imagine when, while sliding an object off a table, the object would begin to fall. For objects with a flat base, once the center of gravity is past the edge of the table, the object falls. So, I can move my coffee mug to almost halfway off the table, but not past that! Another way to visualize this concept is thinking back to the game where you try to balance a bat or a broom on one finger. When the bat is vertical, the center of gravity is over your finger, and the bat can be balanced. But, as the bat falls, the center of gravity moves farther and farther away (since it's around the center of the bat when the bat is horizontal), and therefore the object will fall. Humans can be similarly construed, so when our center of gravity is too far away from our base of support (our feet), we are unstable and will begin to fall.

When the vestibular system fails, the consequences can be bad. In zebra fish, for example, a broken vestibular system is lethal. These small fish—adults measure an inch in length—have five distinct horizontal stripes on each side of the body. They live in freshwater streams and ponds in

Southeast Asia, in the shadow of the mighty Himalayas. Due to a transparent body early in life, unique regenerative abilities, a quick generation time, and a cataloged genome, they are a popular species for medical research. Mutant fish, born without a functional vestibular system (specifically utricular otoconia, which we'll explain later), fail several balance tasks. They neglect to keep their eyes steady when tilted, can't swim upright, and when perturbed, they zigzag, roll, and spiral, instead of swimming in a straight line. Furthermore, many bony fish rely on an air-filled pouch called a swim bladder to maintain buoyancy. Fish will swim to the surface and gulp air to control the volume of the bladder, which then decreases the energy needed to float at a specific water depth. Scuba divers employ a very similar strategy, filling their vests with compressed air to decrease the work of swimming. The mutant zebra fish, with no sense of where they are, fail to fill and regulate their swim bladders. Researchers Bruce Riley and Stephen Moorman concluded that, most likely, the mutant larval zebra fish don't survive until adulthood because without a vestibular system, they cannot swim, float, or navigate, and therefore starve to death.[1]

I've organized this book in several sections in order to try to tell different parts of the incredible vestibular story. In the first section, we'll learn how we evolved the vestibular system, based on the amazing ability of natural selection to adapt biology for remarkably specialized functions. We will also review the history of past scientists and their experiments to try to understand the vestibular system. We'll cover Prosper Ménière's controversial presentation to the French Academy of Medicine in 1861, Robert Bárány's 1914 Nobel Prize in physiology, and what we've learned from J. Richard Ewald's experiments on the inner ears of pigeons. In the second section, we'll take a tour to understand the functioning of the vestibular system, starting with the incredible ability of the hair cell to sense miniscule vibrations, and building up to an understanding of how the brain uses that ability to control eye movements, body movements, and balance. Along the way, we'll meet some fascinating scientists who are studying the vestibular system. We'll learn about their research into understanding the complex web of connections between the vestibular system and other parts of the brain that are responsible for steadying vision, spatial reasoning, and memory. In the third section, we'll learn about different diseases that affect the vestibular system. We'll find out why someone would hear their eyes moving or get dizzy with loud sounds, what happens when crystals run loose in the inner ear, and the consequences when our vestibular system fails to function. Finally, we will pontificate about the future and discuss exciting advances like prosthetic implants that can restore a sense of balance, and gene therapies to rebuild the cellular structures needed for balance.

Before we jump in, it's worth spending a second on nomenclature. In order to discuss the critical concepts of this book, we do need to wade into the jargon of the field. But—fear not—there is an illustrated glossary at the end of the book for quick reference. I tried to keep the writing light and fun, but it wouldn't be possible to discuss some of the more interesting parts of the vestibular system without getting a little technical.

When we discuss the ear, the natural inclination is to picture the part of the ear that is visible. You know that springy, oddly shaped appendage that we sport on each side of our head. In precise anatomical terms, that part of the ear is called the "outer ear." It's also called the "auricle," or the "pinna." The outer ear also includes the ear canal, which is a tunnel that begins with an opening in the auricle/pinna and terminates in the eardrum. Beyond the eardrum lies the "middle ear." The middle ear is a space, bounded on one side by the eardrum and surrounded otherwise by bone. However, the space is filled with air, and just like a balloon, there is also one way in or out. That valve for the middle ear is called the "Eustachian tube," and it's another tunnel that connects to the back of the nose (technically the "nasopharynx"). If the Eustachian tube can't periodically open and equalize, then the middle ear space starts to collapse due to negative pressure. The eardrum can be sucked inward, similar to what would happen if you tried to suck the air out of a water bottle and the plastic sidewalls of the bottle start crinkling inward. Like most bodily systems, equilibrium is key, and interestingly we see a different problem when the Eustachian tube is too open. When this occurs, sounds from inside the throat, like breathing or talking or chewing, get into the ear too easily, where they reverberate, and are therefore loud and distracting.

But back to the nomenclature! The other structures in the middle ear are the ear bones, otherwise known as the "ossicular chain." "Ossicle" is a technical term for ear bone. There are three, named for the everyday objects they resemble. So, we have the hammer, anvil, and stirrup; corresponding to the malleus, incus, and stapes. Now, in my honest opinion, the malleus does not resemble a hammer, the incus somewhat resembles an anvil, and the stapes really resembles a stirrup. But I can't think of an object that the malleus more closely resembles. Finally, we have the "inner ear," which is located further inside the skull. Unlike the middle ear, which is filled with air, the inner ear is filled with fluid. And it's completely encased within bone. The inner ear contains several structures, all of which are connected. The cochlea is the organ of hearing. The semicircular canals and the vestibule are part of the vestibular system. We'll go into more detail in later chapters about all of this.

A disclaimer: There's no perfect way to tell any story. I have tried to present a panoramic view of the vestibular system. To do so, I had to step outside my comfort zone of clinical medicine and venture into strange new worlds. I am certain there are key people, details, and facts that I overlooked, neglected, and downright got wrong. I apologize in advance.

Toward the end of the writing process, I interviewed a number of colleagues, scientists, and experts. They included Joel Goebel, Bryan Ward, Yuri Agrawal, Charley Della Santina, Michael Halmagyi, Taha Jan, and Tim Hullar. I am indebted to them for sharing their wisdom and stories with me.

I am also fortunate to include several images of the microscopic structures of the inner ear, courtesy of scientist Nicolas Grillet. These images are by far the best I've ever seen and open a window into the hidden world of hair cells. They aren't easy to obtain, but luckily for us Nicolas is a master of scanning electron microscopy—a meticulous art that takes years to hone. Nicolas was born in Villeneuve-lez-Avignon, France, and is currently a professor at Stanford University, where he uses mouse genetics to understand the biology of the inner ear on a deep and fundamental level.

A few words on my teaching philosophy during the writing of this book. There are different levels of explanation for any topic, so please indulge my choices. Too zoomed out, and our view would be too blurry to appreciate important details and concepts. Too zoomed in, and we would lose the forest for the trees. Volumes could be written on each topic covered by a single chapter. PubMed, a digital directory of published scientific research, currently references thirty million manuscripts. I tried to keep things interesting, accurate, and relevant. Others have written textbooks, but lacking their attention span, I instead wanted to curate a scientific tour, fueled by curiosity. I hope you enjoy.

PART 1

HOW WE GOT HERE

1

An Aural History

Let's start with some background. The structures in the human ear are among the tiniest in the human body. The stapes bone, measuring about four millimeters in height, is the smallest bone in the human body. For comparison, that's about the size of a grain of rice. In fact, it was not "discovered" until the mid-sixteenth century. Despite its diminutive size, it sits like a giant atop the inner ear, whose structures are microscopic. The entire inner ear contains about 200 cubic millimeters of fluid, about the same as four raindrops.[1] The width across each semicircular canal is about 1 millimeter, or roughly the width of a pencil tip. And, to complicate matters further, the entire structure is housed within the densest bone in the human body. We call this bone the otic capsule because it serves to protect the delicate mechanism of the ear (otikos is ear in Greek). Therefore, it's not entirely surprising that the function of the inner ear was not discovered until relatively late in history.

The physicians of yore, informed by fanciful theories of health and disease, offered various remedies for vertigo. One fine fall day, while trying to pass time in Vancouver, I found myself in the rare book section of the city library. I inquired about old medical texts and was rewarded with a copy of Nicholas Culpeper's *The English Physician*. Published in 1708, it reads like a magical potion book, with instructions for "a Complete Method of Physick, whereby a Man may preserve his Body in Health, or Cure himself, being

Sick, for Three Pence Charge, with such things only as grow in England, they being most fit for English bodies." Culpeper was a master of astrological botany, advising that herbs had to be harvested at proper times according to the position of celestial bodies, ensuring proper pairings between planetary personalities and medicinal concoctions.

Culpeper begins with a description of Amara Dulcis, also known as bittersweet, woody nightshade, and felonwort. He writes, "It is under the Planet Mercury, and a notable herb of his also, if it be rightly gathered under his influence. It is excellent good to remove witchcrafts both in men and beasts; as also all sudden diseases whatsoever. Being tied round the neck, is one of the admirable remedies for the vertigo or dizziness in the head that is; and that's the reason the people in Germany commonly hang it about their cattle's neck when they fear that any such evil hath betided them."[2] Of note, just to present an opposing point of view, Wikipedia states, "Solanum dulcamara has a variety of documented medicinal uses, all of which are advised to be approached with proper caution as the entirety of the plant is considered to be poisonous." Before you judge Culpeper too harshly, consider that astrology and signs of the zodiac are still immensely popular today.

Starting in the sixteenth century, Italian anatomists began to dissect and understand the internal structures of the ear in detail. In fact, much of the anatomy in the ear is named after these Italian scientists. The Fallopian canal is a twisty tunnel that traverses the temporal bone. It contains the facial nerve, which is the nerve that carries the signals from our brain to the muscles in our face. The face muscles are then activated for facial expression, with different patterns depending on whether you want to smile, or raise your eyebrows quizzically, or flare your nostrils. Interestingly, the muscles of the face are some of the only muscles that connect bone to skin. Most other muscles connect bone to bone, so that we can move our bones relative to each other (like, say, to walk). The purpose of having muscles connected to skin is mostly so that we can express ourselves emotionally, by changing the shape of our face. It's interesting to think about the evolution of that ability, which is not present in most animals (and may explain why we don't find fish to be very emotive). Every budding young ear surgeon must dedicate hundreds of hours of cadaveric study to learn the course of the Fallopian canal, because the consequences of damage to the facial nerve—a paralyzed and drooping face incapable of expression—is devastating, and likely also a lawsuit. That canal is named after Gabriele Falloppio. You may also recognize his name in the *other* Fallopian structure, the *tube*, which connects the uterus to the ovaries. He also named the cochlea, noting that its spiral resembled a snail's shell, which is the Latin root of the word.

Bartolomeo Eustachi was another important Italian anatomist. His name lives on in the Eustachian tube, which is the valve that allows pressure equalization for the middle ear. He also described the two tiny muscles that are in the middle ear: the stapedius and the tensor tympani. The function of those two muscles is to protect our ears. Have you ever been told not to stare into the sun? Similar to the visual system, where incredibly bright lights can cause retinal damage, loud sounds can damage the organ of hearing, the cochlea. The technical term for this is called excitotoxicity. So, when we see something very bright, we instinctively close our eyes, to protect them. But how do we close our ears? One mechanism—which is quicker than putting our fingers in our ears—is for these miniscule muscles to contract. By pulling on the ossicular chain, they stiffen it. That dampens sound vibrations, so that the noise level of the sound reaching the cochlea is reduced. Eustachi was a contemporary of Andreas Vesalius, who is credited as a grandfather of human anatomy.

In 1772 and 1789, Antonio Scarpa published works on the anatomy of the inner ear. He recognized that the inner ear had two parts, separated by a membrane. By way of analogy, let's picture a pool, and you've tossed a water balloon inside. And to make things clearer, in order to needle you, your brother has added some yellow dye to the water inside the balloon. Perhaps

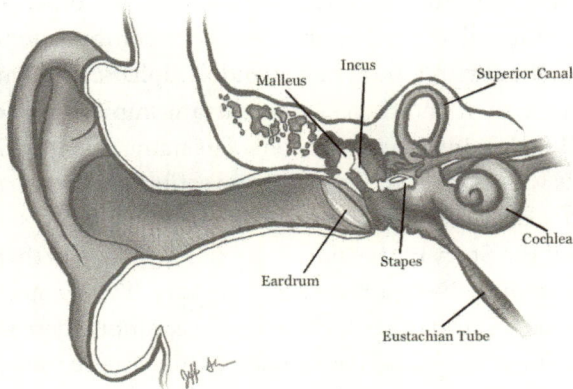

FIGURE 1.1 Anatomy of the ear.

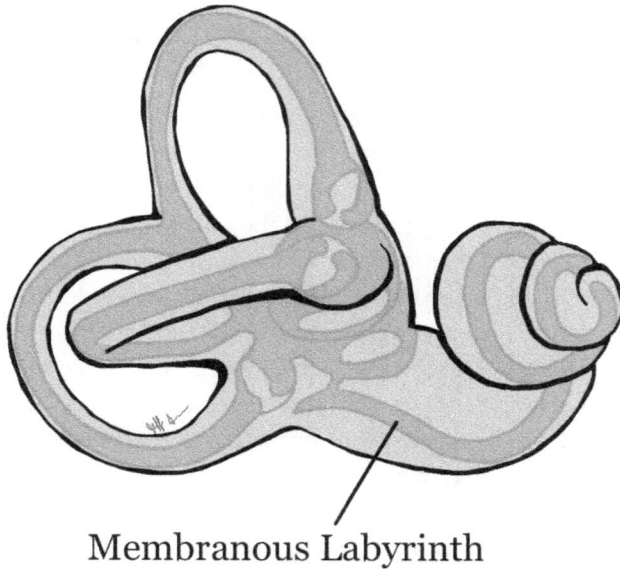

Membranous Labyrinth

FIGURE 1.2 The membranous labyrinth.

he wanted you to assume that the water in the balloon was a different sub-stance! At any rate, now we have a pool, surrounded by concrete, with two compartments separated by the balloon: inside the pool but outside the balloon, and inside the balloon. Similarly, the inner ear has two compart-ments separated by a membrane: inside the bone but outside the membrane, and inside the membrane. The membrane is called the membranous laby-rinth (and its shape is quite complex, like a maze). The outer component is filled with a fluid called perilymph. The inner compartment (inside the membrane) is filled with a fluid called endolymph. Scarpa was one of the first to recognize that, and there are important implications of the struc-ture that we'll review in coming chapters. His name is lent to Scarpa's gan-glion, which is where the cell bodies of the vestibular nerve are located. Of note, you can visit Scarpa to this day! His preserved head is on display at the Museum of the Story of the University of Pavia, in Northern Italy.

So, on approaching the mid-nineteenth century, it was generally accepted that the inner ear was necessary for hearing. In addition, there was anatomic proof that the inner ear had other structures, like the semicircular canals, although their function was not known. That began to change with Flou-rens. Born in 1794, Jean Pierre Flourens was a French scientist who studied the function of different parts of the brain. His experimental approach was to lesion (i.e., destroy) a specific part of the brain, and then observe what changes in behavior could be observed. For example, he found that if he removed the cerebellum, animals would lose coordination and equilibrium. So, he surmised that the cerebellum must have a role in the coordination of muscle activity (which is true!). He found that if he damaged the brain stem, animals would die, leading him to deduce that the brain stem was integral for the maintenance of life (also true!). Flourens provided some of the earliest insights into the function of the semicircular canals. Keep in mind that we have three semicircular canals in the inner ear on each side of our head (or six in total). They are each located in a different plane, so as to provide information about head movement in any possible direction. Otherwise stated, the reason that we have three semicircular canals is the same reason that there are three axes in any coordinate system. From his experiments, "If the horizontal canals are severed, the animal turns on its vertical axis; if the posterior vertical canal is severed the animal rolls over backward, and if the anterior vertical canal is severed the animal falls forward."[3] So, specific damage to different parts of the inner ear would cause a different pattern of turning behavior in the animal, in the same plane as the canal being damaged. While the full implications of that finding were not yet clear (Flourens thought that the function of the semicircular canals was to inhibit motion), this was a huge step forward. Flourens is also credited with establishing that only the cochlea, and not the other parts of the inner ear, are necessary for hearing. This pioneering work was cited by both Ménière and Bárány, and as is often the case in science, helped give them a foundation from which they were able to expand and broaden knowledge.

In 1861, Prosper Ménière, a Parisian neurologist, presented his research to the French Academy of Medicine.[4] At the time, he was the head of the Imperial Institute for Deaf-Mutes. He presented a series of cases to support an argument that vertigo could arise from the inner ear. The prevailing orthodoxy held that vertigo was a symptom of "cerebral congestion," which was typically treated with bloodletting. Ménière presented autopsy findings from a young patient who suffered from vertigo and hearing loss. He found that the brain and spinal cord appeared normal, but that the inner was filled with a bloody exudate. This inner ear damage, he argued, was the cause of the girl's vertigo and hearing loss. He also described a case of a man with repeated episodes of vertigo and hearing loss, which first affected one ear, and then the other. Based on the fact that the hearing loss and vertigo would occur together, he surmised that they were caused by the same disease process, affecting the inner ear and not the brain. Ménière's intuition was correct. We now know that because the vestibular system and the hearing system occupy the same space (like roommates), diseases often affect both systems (like, say, a leaky tub that causes water damage to both roommates). So labyrinthitis, perilymph fistula, temporal bone fracture, and superior canal dehiscence can all cause hearing and vestibular damage. Ménière's name lives on Ménière's disease, which causes hearing loss, tinnitus, ear pressure, and vertigo. It's sobering to reflect that over 150 years after the initial description of the disease, we still do not know what causes it.

THE PIGEONHOLE

Born in 1855 in Berlin, J. Richard Ewald was a German physiologist. As the chair of physiology in Strasbourg, he is best known for his experiments on pigeons and compiled his research into a landmark text *Physiologische Untersuchungen Ueber Das Endorgan Der Nervus Octavius*, published in 1892. This translates to "Physiologic Studies on the End Organ of the Eighth Nerve." Now, geography buffs (and I hope you are out there!) might point out that Strasbourg today is part of France, not Germany. Located on the border between the two countries in Alsace, it's changed hands frequently throughout history. It was captured by Germany in 1871 during the Franco-Prussian War, and it would remain German until the end of World War One, in 1918. So, during Ewald's time, it was a German city. It was again captured by Germany during World War Two, only to be reunited again with France in 1944.

The nerves from our brain that go to different parts of our head (and not the rest of the body) are called cranial nerves, and they are ordered according to where they leave the brain, from I (in the front), to XII (in the back). So, the first cranial nerve allows us to smell, while the twelfth moves our tongues. The eighth nerve handles hearing and balance. End organ refers to the thing that the nerve is plugged into. So, for the second nerve, which handles sight, the end organ would be the eye (and more specifically the back part of the eye that senses light, which is the retina). The eighth nerve has several end organs, which include the cochlea (hearing), the semicircular canals (balance), and the otolith organs (balance).

In his experimental setup, Ewald performed surgery on awake pigeons. He would make two small openings into a semicircular canal and then plug the inside of one. With the other opening, a pneumatic hammer was introduced, which could gently squeeze the membrane inside via a handheld bulb.

Since the semicircular canals are normally activated by head movements that cause fluid to move in the inner ears (similar to the way orange juice hits the walls of the container as you shake it), Ewald's hammer was an experimental way to mimic the natural activation of the semicircular canals. The plug was used to direct the force toward the part of the semicircular canal that is the sensor (which is called the ampulla). With these experiments, Ewald formulated three laws of vestibular physiology. These laws are foundational, so I include them in every vestibular teaching lecture I give.

Ewald's first law states that evoked eye movements will be in the plane of the canal being stimulated. So, if the horizontal canal was stimulated, the eye movement would be horizontal, and if the vertical canals were stimulated, the eye movements were vertical. This can be thought of like a marionette doll, where the motion of the body and limbs are controlled by strings attached to a control bar. If you tilt the bar up, the right hand also moves up. Since they are directly connected, the movement is in the same plane. Remarkably, the vestibular system exerts similar control over the eyes (although there are no tiny strings in our head—just neural connections). This ends up being a fascinating example of form following function. A primary purpose of the semicircular canals is to react each time the head moves and keep the eyes steady with a countermovement. With their anatomic structure, information about head movements is automatically parsed into signals regarding movement in each possible vector. Because the eye movement is compensatory for the head movement (head looks to the left, eyes then compensate by moving to the right so that overall, there is no net movement), it would make sense that the eye movement should be in the exact same plane as the head movement. That way, the brain does not need to perform any calculations to determine the proper compensatory eye movement, as the necessary information has been hardwired in the system. We'll come back to this idea in chapter 4.

Ewald's second and third laws concern which specific stimulus excites or inhibits each canal. To understand, we'll have to introduce a little more anatomy. Each semicircular canal has one portion that acts as the sensor. The rest of the canal is there to ensure that the inner ear fluid has a pathway to move about, so that moving fluid in the canal can either push or pull on the sensor. Ewald's second law states that for the horizontal canal,

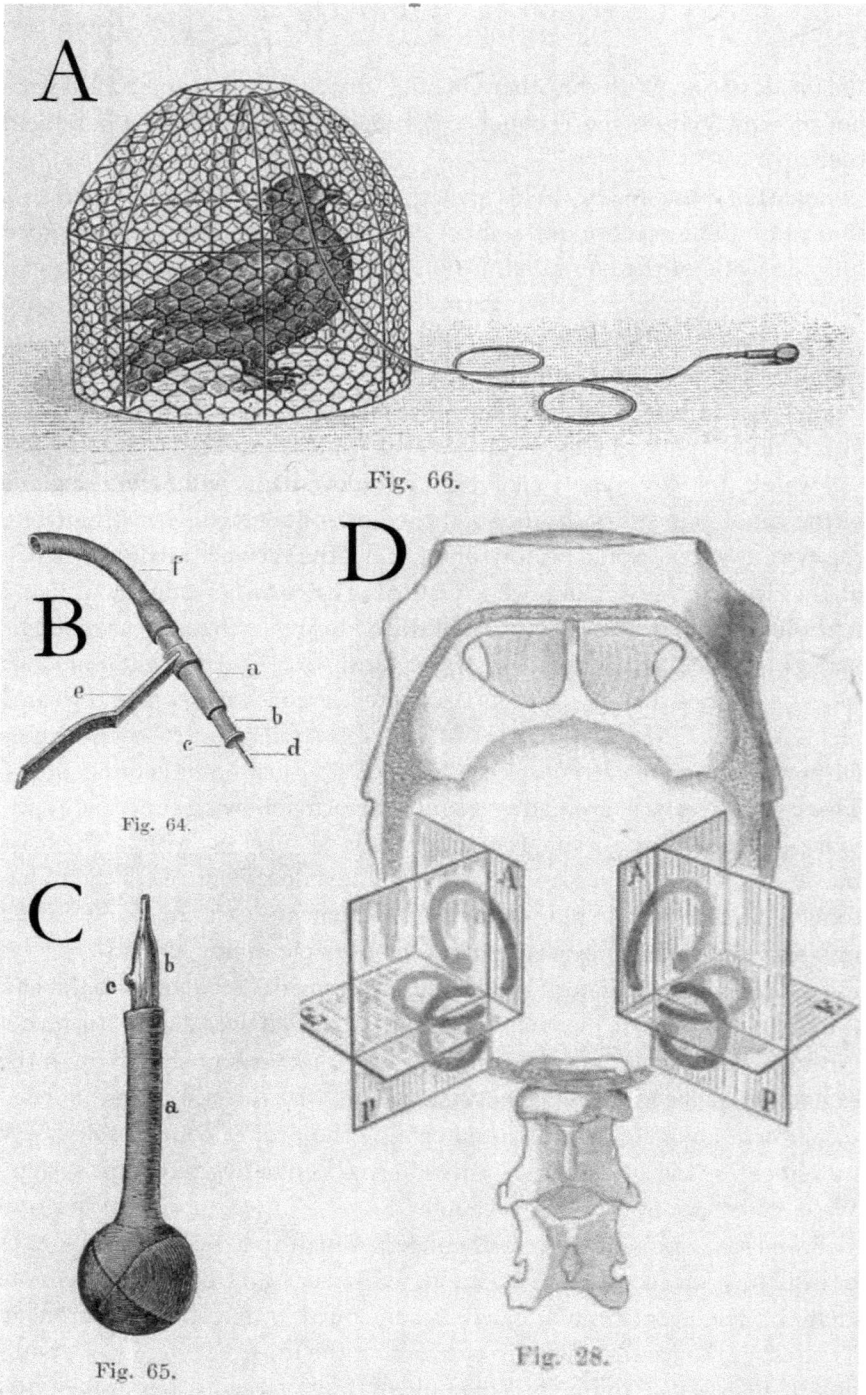

FIGURE 1.3 Ewald's experimental setup: (*A*) the pigeon and the attached pneumatic hammer; (*B and C*) details of the pneumatic hammer; and (*D*) the three planes of the vestibular system. (Courtesy of Archive.org)

"ampullopetal" flow is excitatory (the sensor being pushed, from the standpoint of the canal), and the third law states that for the vertical canals, "ampullofugal" flow is excitatory (the sensor being pulled, from the standpoint of the canals). The ampulla refers to the widened portion of the semicircular canal that contains the sensor. You can recognize the Latin roots "petere," meaning to seek, and "fugere," meaning to flee. Fugere is in many modern words, like "fugitive," or one who flees. The cupula is within the ampulla. It's like the weathervane of the inner ear and will deflect back and forth depending on the movement of the fluids. If it moves toward the vestibule (the central room of the inner ear), the movement is described as ampullopetal. If it moves away from the vestibule, then it's ampullofugal. So, different directions of cupular movement are either excitatory (more nerve firing), or inhibitory (less nerve firing), depending on the canal that they are in. We now know the reasoning behind that feature that Ewald discovered, and it's related to the how the hair cells are oriented (to be reviewed in Part II). While it's impossible to say exactly why we are wired this way, one thing to note is that from the standpoint of our heads, any movement in the plane of a canal toward that side (right or left) is excitatory. So, the principle behind Ewald's second and third laws may relate to the advantage of each canal providing maximum, non-overlapping information.

In his seminal text, Ewald also described many other experiments on pigeons and other animals. Using a camera lucida, he captured the behavioral changes of the animals after removal of the inner ears. For instance, he notes that after labyrinthectomy, his pigeons could not handle a twenty-gram lead ball attached to their beak. According to Google Translate (did I mention that I don't understand German?) he writes, "But how different the picture is with the labyrinthectomized pigeon! For them the ball is a tremendous burden." So, I guess I do now know one German phrase, which is "labyrinthlosen Taube"—aka the labyrinth-less pigeon. I feel that this will come in handy one day.

A DISCOVERY OF COWS

In 1914, Robert Bárány was awarded the Nobel Prize in physiology for his work on the vestibular system. His acceptance speech provides a remarkable insight into the story behind his discoveries, which today still form the basis for laboratory testing of the vestibular system, and our general

Fig. 2.

Fig. 11.

Fig. 50.

FIGURE 1.4 Labyrinthectomized animals: (*A*) the Labyrinthlosen Taube struggles with a lead ball; (*B*) labyrinthectomized pigeon having an unconventional drink; and (*C*) labyrinthectomized frog turning its head. (Courtesy of Archive.org)

understanding of vestibular physiology. His account is reproduced below (from nobelprize.org):

As a young otologist I worked in Professor Politzer's Clinic in Vienna. Among my patients there were many who required syringing of the ears. [This practice, still done today, refers to cleaning out earwax by using a syringe filled with water to flush out the wax] A number of them complained afterwards of vertigo. Obviously I examined their eyes and I noticed in doing this that there was nystagmus in a certain direction.

Nystagmus, as we will discuss in detail later, is a rhythmic back-and-forth movement of the eyes. So it looks like a jiggling of the eyes. Bárány continues:

I made a note of this. After a time, when I had collected about twenty of these observations, I compared them one with another and was amazed always to find the same note. I then realized that some general principle must be implied, but at the time I did not understand it. Chance came to my aid. One of my patients, whose ears I was syringing, said to me: "Doctor, I only get giddy when the water is not warm enough. When I do my own ears at home and use warm enough water I never get giddy." I then called the nurse and asked her to get me warmer water for the syringe. She maintained that it was already warm enough. I replied that if the patient found it too cold we should conform to his wish. The next time she brought me very hot water in the bowl. When I syringed the patient's ear he shouted: "But Doctor, this water is much too hot and now I am giddy again." I quickly observed his eyes and noticed that the nystagmus was in an exactly opposite direction from the previous one when cold water had been used. It came to me then in a flash that obviously the temperature of the water was responsible for the nystagmus. From this I immediately drew certain conclusions. If the temperature of the water was really responsible, then water at exactly body temperature should cause neither nystagmus nor vertigo. An experiment confirmed this conclusion.

It's worth emphasizing a few points. First, Bárány stumbled on a remarkable discovery somewhat by chance, but also by paying close attention to his patients. Second, his patient had really noticed the phenomena of cold water causing vertigo, but Bárány was diligent and curious enough to follow up on his observations. Bárány then continues to describe the implications of his hypothesis, and how he tested the predictions that logically followed from his ideas.

"Furthermore, I said to myself, if it is the temperature of the water, nystagmus must be caused in normal cases also and not only in cases of suppurating ears." Suppuration of the ears refers to an ear that has been permanently damaged due to an infection, with pus inside the ear. Bárány is arguing that if the nystagmus depends on a functioning inner ear, then if the inner ear is not functional—due to a prior infection—then no nystagmus should be caused with syringing. Of note, ear infections that cause vestibular damage are thankfully rare in our times, due to antibiotics, vaccines, and ear tubes. Bárány goes on:

> This I was also able to prove. Because of my earlier research, I did not doubt for a second that the nystagmus was the result of a reflex action of the semi-circular canals. Hence, the further conclusion followed that if these were destroyed there would be no reflex action. I was able now to look among the abundant material available in the Vienna Otological Clinic for a suitable case. Before long, I found a case of severe suppuration of the middle ear, in which, even after continuous cold syringing, there was no nystagmus reaction.

Having confirmed his theory that nystagmus—the rhythmic eye movement—was dependent on the inner ear, Bárány then explains how he figured out the mechanism behind this reaction. He wanted to understand why cold water in the ear canal caused the nystagmus in one direction, and warm water caused nystagmus in another direction:

> I had already recognized the importance of the caloric reaction and yet I could not explain it. In vain I reflected upon it. Then, one day, I had an idea. I remembered the bath water tank and my surprise, as a child, at finding the water immediately above the fire quite cold, whereas higher up, at the top, the tank was so hot that it burned one's finger. The labyrinth reminded me of a bath-water tank, i.e. a container filled with fluid. The temperature of the fluid is, of course, 37°C— body temperature. Suppose I spray one side of the container with cold water? What

will happen? The water on this side of the tank will cool, of course, and, therefore, it will attain a higher specific weight than the surrounding water and will sink to the bottom of the container. Other water at body temperature will take its place. If I syringe the ear with hot water, on the other hand, the movement will be exactly the opposite. The movement of the fluid must change, however, when I alter the position of the container, and if I turn the container through 180°, it must change in exactly the opposite direction. Immediately I was able to envisage the kind of test which would serve as experimentum crucis for this theory. If, in two head positions differing from one another by 180°, it is possible to obtain nystagmus in opposite directions by syringing, whether with hot or cold fluid, then this theory must be correct. I went to the clinic and arranged the experiment and, in fact, the hoped-for result showed itself very clearly. Two head positions differing by 180° show nystagmus in exactly opposite directions. The theory of the reaction was now established and it agreed absolutely with the theory of both Breuer and Mach which had recognized the movements of the endolymph, the fluid contained in the semi-circular canals, as being the cause of stimulation in them.

You may recognize the two names mentioned by Bárány. Both Joseph Breuer and Ernst Mach are famous in their own right and it's not for their seminal contributions to vestibular physiology. Joseph Breuer was Sigmund Freud's mentor and developed the "cathartic method" of psychoanalysis after noting that a patient's symptoms radically improved after discussing them. Ernst Mach described and photographed shock waves of objects moving faster than the speed of sound, and now his name is forever enshrined with the Mach number. Independently, in the 1870s, both asserted that endolymph movement in the semicircular canals was the foundation of vestibular sensation.[5] It seems that Mach's interest in the vestibular system was sparked by a keen observation:

My view at that time was that the whole body contributed to movement sensation. The supposition of a special organ for movement sensation was far away at that time. . . . A chance happening led me back to the study of motion sensation. I noticed the tilting of houses and trees while I was travelling around a curve in a railroad. This is easily explained if one directly senses the resultant inertial acceleration.[6]

Mach followed up with a number of experiments where he spun subjects around in a chair on various axes, frequently in the dark. The subjects would report how they felt they were moving, which was compared with how they were actually moving. By analyzing the differences between perception and reality, Mach was able to deduce important principles of vestibular sensation. For example, subjects rotated on a chair in the dark would correctly report that they were spinning, but that sensation decayed after a few seconds. So, twenty seconds into a spin in the dark, his subjects couldn't tell they were spinning anymore. Furthermore, when the chair was stopped, subjects felt they were spinning in the opposite direction. This led Mach to conclude that "one therefore doesn't sense angular velocity, but rather angular acceleration." Mach and Breuer differed slightly in their view of the endolymph force. Breuer believed that endolymph fluid flow was sensed by the semicircular canals, whereas Mach believed that the fluid didn't move, but instead exerted pressure due to inertia.

Breuer's contributions were remarkable, and helped lay the foundation of a modern understanding of vestibular physiology. He discovered much of what we'll cover in chapter 4: the mechanism behind the semicircular canal's ability to sense head turns. He helped decipher nystagmus— including the slow and quick phases—and the function of the otolith organs. He confirmed his theories regarding the sensory function of the vestibular system with animal experiments, by dissecting out the miniscule nerves to each canal, and then paralyzing them with cocaine crystals. (Author's note: cocaine is the original topical anesthetic, which is why all the synthetic derivatives that we use today—like lidocaine or prilocaine— sound similar. They work by blocking nerve signals, and that causes numbness because the sensory nerves that normally signal pain are blocked. I've been told by friends that cocaine has other, nonmedical uses as well). The cocaine paralysis in each canal produced nystagmus in the plane of the tested canal, proving Breuer's theories to be true. However, perhaps even more remarkable was that Breuer didn't have a university appointment. As relayed by Doctors Gerald Wiest and Robert Baloh, Breuer's hundreds of animal experiments were conducted in his tiny apartment in Vienna.[7] Breuer wrote, "I am a practicing physician. My time for experimentation is limited to the late evening and the night, and my laboratory is my home."

Today, in medical schools around the world, young physicians are taught the acronym "COWS." It stands for "cold opposite, warm same." They are being asked to memorize the results of Bárány's observations. The eye jiggling—nystagmus—that results from the vestibular system is a back-and-forth movement. But the movement in one direction is faster than the movement in the opposite direction. In gym class, we used to have races where you would first run facing forward, but then you had to get back to the starting point racing backward. If you had to run several laps of that, you would see a similar pattern to vestibular nystagmus, where each time the runner was going in the forward direction, the speed was "fast," and each time they were running backward (all on the same track), the speed was "slow." So, when you look at an eye with vestibular nystagmus, you'll see a quick phase, and a slow phase, and they repeat (slow-fast-slow-fast-slow-fast . . .) until the nystagmus stops (usually a minute or two with Bárány's method). By convention, the nystagmus is named for the fast phase. So, if the direction of the fast phase is to the right, then we call it right beating nystagmus. What Bárány saw is that if you lie on your back, and you put cold water in the left ear, the nystagmus will beat to the right (i.e., the fast phase is to the right). And he was mostly right about the mechanism. The cold water increases the density of the endolymph—the fluid in the semicircular canal. So, the fluid will move downward. For reasons we'll discuss later that fluid movement causes the same signal in the vestibular nerve as if the head were being turned to the right. With a head turn to the right, the vestibular system needs to compensate and move the eyes to the left. The result is no net movement, so whatever you are looking at stays steady in your vision. But eyes cannot continue to rotate in the eye socket indefinitely (unlike cartoon characters!). So, we protect ourselves with a quick resetting eye movement. This is similar to when a typewriter resets at the end of a line of text. This quick resetting movement is the fast phase.

Bárány had figured out how to use temperature to stimulate the vestibular system. He also introduced another, more direct method: The Bárány Chair. Since the vestibular system normally responds to rotations, it's logical to use rotations to study the vestibular system. Using a custom chair, subjects would be rapidly whirled about. Sensing this, the horizontal canals tell the eyes to react, so that a spin to the right causes an eye movement to the left. Similar to the caloric response detailed above, sustained rotation causes nystagmus. The reflexive eye movement to the right is reset each time with a quick leftward correction, in a back-and-forth dance of the eyes. By comparing how long the nystagmus lasted after a clockwise vs. counterclockwise spin, Bárány would deduce if the right or left vestibular system was damaged. While Bárány brought the rotary chair into the field of

FIGURE 1.5 William Saunders Hallaran, "Practical Observations on the Causes and Cure of Insanity" (Cork, 1818). (Courtesy of Wikimedia Commons)

vestibular medicine, it was already a popular treatment in the nineteenth century. At that time, many prominent physicians throughout Europe believed that rapid rotation was an effective treatment for insanity. Illustration 1.5 is from a psychiatry textbook written by William Saunders Hallaran, director of the City of Cork Lunatic Asylum. Therefore, the historical record points to Dr. Hallaran as being the original "spin doctor" (forgive me dear reader—I simply could not resist). I will admit, viewing the poor soul strapped to the rotating box, and the stooped physician twirling him, it's tough to tell who the insane one is. That sentiment lasts to this day, as we do use a modern version of the Bárány chair to study vestibular responses and frequently receive quizzical looks from our patients as they are harnessed into a mechanized, medical Tilt-A-Whirl.

The influence of the incredible people we met in this chapter lives on. Bárány gives his name to the Bárány Society, which is an international group of vestibular researchers. Neuroscientists are taught the Laws of Ewald, governing the basic principles of the vestibular system. And Ménière's disease continues to be discussed every day, as scientists around the world race to finish his story, by finding a cure to that horrible disease.

2

How Cheetahs Prosper (Evolution)

CELL FUN

Let's set the stage. The universe is about thirteen billion years old, and Earth about 4.5 billion years old. Life first appeared with living creatures that were just a single cell. A cell is a microscopic being, with a few essential parts needed to eat, reproduce, and defend itself. And if you think about all the bacteria around today, you'll see that even a single-celled organism can be quite formidable. At some point along the way (hundreds of millions of years ago), life forms evolved into assemblages of cells. In so doing, these multicellular creatures exploited the advantage of having larger bodies, where individual cells are adapted to a variety of functions. Humans, for reference, have tens of trillions (!) of cells in our body. And even though it's all just cells, some are superspecialized. Skin cells, interleaved into watertight barriers, envelop and protect our bodies. Muscle cells are arranged in rows like the pistons of a car engine, shortening (contracting) to power movement. And nerve cells, or neurons, are organized into a massive biologic computer capable of contemplating the quantum world of the atom and the colossal vastness of the cosmos.

How did this all come to be? How did life evolve from organisms smaller than a fleck of sand into the mighty elephant? Or the mighty duck? The fundamental process that governs the change in life forms over time is natural selection. Similar to the way software code instructs a 3D printer to create an object, DNA guides the assembly, structure, function, and arrangement of our cells. If you want to truly be amazed, just spend some time contemplating the fact that we self-assemble into a fantastically complex structure—capable of self-awareness—from a pairing of just two cells.

One of my favorite short stories was written by Jorge Luis Borges, an Argentinian author active in the mid-twentieth century. In "The Library of Babel," he describes a surreal world occupied by an infinite library and its inhabitants. The books of this library have been assembled through random permutations of all the letters of the alphabet and punctuation marks. Therefore, most of the writings are utter nonsense, just random strings of letter combinations. As you can imagine, this preponderance of gibberish takes an emotional toll on the inhabitants, who search for meaning by combing the endless stacks for years. However, as Borges writes, the library also contains every single book that you could possibly imagine:

> Everything: the minutely detailed history of the future, the archangels' autobiographies, the faithful catalogues of the Library, thousands and thousands of false catalogues, the demonstration of the fallacy of those catalogues, the demonstration of the fallacy of the true catalogue, the Gnostic gospel of Basilides, the commentary on that gospel, the commentary on the commentary on that gospel, the true story of your death, the translation of every book in all languages, the interpolations of every book in all books.[1]

Given the power of infinity, even a truly random process will produce highly ordered entities. Our genetic code, subject to random mutations over eons, is an engine of innovation introducing small changes into our assembly instructions. Most of these changes are not helpful, like the vast majority of the books in the Library of Babel. However, the library—and our genetic code—also contain the necessary blueprints for assembling an eye. While time is not infinite, several billion years was clearly an adequate epoch for interesting things to happen. Luckily, natural selection is a goal-directed process, unlike Borges's library, and changes that result in betterment of the organism (increased fitness to survive and reproduce) will result in increased proportion of those traits in subsequent generations. So, with countless iterations of this process, errors in the transmission of genetic language that are helpful toward reproduction (which is the only measure in natural selection, but obviously survival is a prerequisite) will accrue. These adaptations, like the vestibular system, exploit our physical environment on Earth. If you can think back to the fundamental properties of physics, there is an animal on Earth who has likely taken advantage of that property because of natural selection. We humans can see light, which is a form of electromagnetic radiation. We can also hear sound waves, which are emitted pressure disturbances that shake air molecules as different things collide. Sharks can sense electricity, and some migrating birds and marine mammals appear to be able to sense Earth's magnetic field. In his landmark text, *The Origin of Species*, Charles Darwin wrote one of the most inspiring and illuminating sentences in all of human history:

> Thus, from the war of nature, from famine and death, the most exalted object which we are capable of conceiving, namely, the production of the higher animals, directly follows. There is grandeur in this view of life, with its several powers, having been originally breathed into a few forms or into one; and that, whilst this planet has gone cycling on according to the fixed law of gravity, from so simple a beginning endless forms most beautiful and most wonderful have been, and are being, evolved.[2]

CENTERS OF GRAVITY

The earliest "vestibular" ability that appears in living things is the ability to sense gravity. Plants, snails, eagles, lemurs, eels, and humans are all able to do so. Keeping in mind that the simplest solution is often the best, the simplest method of finding the vector of gravity is a plumb line. A plumb line is a string attached to a weight. When the weight is dropped, it will align vertically, as gravity pulls the weight downward. That basic concept, of a relatively heavier weighted object, having freedom of movement and therefore resting in the most gravity-dependent position, is used by all the living things mentioned above to "sense" gravity. However, the exact mechanism does differ by the type of plant or animal. Plants have a "statocyte"—a specialized cell—which contains a "statolith"—a ball of starch that serves as the weight. Many sea creatures, like snails, squid, and shrimp possess a "statocyst," which is a spherical organ with a mineralized weight inside— also called a "statolith," We'll see the root "lithos" again shortly, in ancient Greek it translates to "stone." No matter what position the animal assumes, the heavy statolith will always roll to the bottom of the statocyst, enabling the animal to literally know "what's up." Humans and other vertebrates use a similar system, where otoliths (miniature crystals) sit atop the utricle on a bed of gelatinous goo, and slide around as the head is moved, resting in the most gravity-dependent position.

The weighted object—statolith or otoconia—is only half the picture. It's also necessary to have a sensor, which in animals takes the form of a hair cell. As we'll discuss in future chapters, the hair cell is the essential piece of biologic technology that allows for hearing and balance. While forms can differ, it consists of a cell with protruding rods. When the rods are bent one way or another, the cell can sense that and send those signals to the brain. In the human inner ear, those rods are called stereocilia. Try this: close your eyes and then place the palm of one hand over the fingertips of the other hand (like the hand position of a basketball player calling for a time out). Your fingers are the stereocilia. Now, move the hand on top back and forth, and side to side. You can tell that the hand is moving because your fingers are being bent. Stereocilia function in the same way; as they are deflected in a direction, little openings are created along their length, and this begins a process of changes in the cell that ultimately result in a nerve signal.

While the prehistoric history is inherently murky, it seems likely that the cochlea and the semicircular canals developed as offshoots of the simple gravity sensing organ.[3] The structure of the inner ear would seem to support this notion, as both the semicircular canals and the cochlea appear to sprout outward from the centrally located vestibule, which houses the gravity sensing otolith organs (called the utricle and the saccule). In fact, this helps explain the name "vestibular." Vestibular derives from the Latin "vestibulum," which translates as "entrance court."

Fundamentally, the structure of the hair cell allows them to sense when their protrusions—the stereocilia—are being stretched in one direction or another. That ability can be used for other purposes, such as sensing head turns or sound waves. And, as we'll discover, fish use that ability for a different purpose unique to the aquatic environment—to sense the water currents around them.

In a 2014 article, Bernd Fritzsch and Hans Straka explain the evolutionary origin of the hair cell.[4] The common ancestor of all animal life is a single-celled organism called a choanoflagellate (which just rolls off the tongue, doesn't it?). The name is actually quite descriptive: the cell's bottom is shaped like a funnel (choano) and there's a tail (flagella) protruding out of that funnel. The tail is thought to have two purposes: movement (picture a crocodile swimming by undulating its tail) and feeding (the tail also acts like a tongue, bringing tasty bacteria to the cell for ingestion). What's important here though, is appearances. The choanoflagellate looks just like a hair cell. Both are composed of a cell body with a protruding appendage: stereocilia (hair cell) or flagella (choanoflagellate). Fritzsch and Straka argue that since the basic design of the hair cell is known to mother nature and universally present in animals, it wasn't hard for her to co-opt the cellular design and repurpose it for gravity sensation.

Many aquatic invertebrates—such as crabs—evolved semicircular canals independently of vertebrates. This is considered to be an example of convergent evolution; whereby different species of animals seem to have evolved a similar solution independent of each other. A classic example of convergent evolution is the shape of wings. A bat—which is a mammal—has a very similarly shaped wing to most birds. Their wings evolved independently of each other, since mammals and birds occupy different branches on the evolutionary tree. However, since there is an ideal shape for flight, both arrived at a similarly efficient solution. The shape of aircraft wings and helicopter rotors also follow the same basic design. Similarly, adding circular offshoots to the inner ear, with the same stretch receptor used to sense gravity, is useful, so that both crabs and humans independently evolved that ability.

Fish use hair cells for another ability that humans are not capable of. They can sense water movements around them. As you can imagine, this is quite useful, as it allows them to detect approaching predators, stay in tight formation in a school, or even chase prey. The "lateral line" is a sensory organ which is visible as a distinct stripe along the side of many fish. Water can freely enter the lateral line through numerous pores and will deflect hair cells dependent on water perturbations. Picture jumping in a pool while your friend is holding a foam noodle. As the ripples extend outward, they will bend the noodle, letting your friend know where you jumped in. Interestingly, the hair cells in the lateral line also increase the area of their sensor with a cupula—which is a jellylike extender. Our semicircular canals also use a cupula, so that as much movement energy as possible is collected and transmitted to the hair cell. Therefore, more accurately, the cupula moves due to surrounding fluid currents (seawater for the fish lateral line, endolymph for our semicircular canals), and the hair cells, embedded in the flexible cupula, bend as they are pulled or pushed by the cupula.

Birds have also found a unique and wondrous use for the hair cell. Just like an airplane, almost all avian species come equipped with an altimeter called the "paratympanic organ."[5] The name describes its location: the paratympanic organ is located in the middle ear, nestled between the eardrum and the bony walls that protect the inner ear. When you consider the vast distances traversed by migratory birds, the utility of sensing altitude makes sense. The arctic tern is famous for its yearly voyage between the North and South Poles. Which, of course, brings up a key question: Are there any flat-earthers within the tern community? The record for the longest nonstop avian flight, however, goes to the bar-tailed godwit. While being tracked by satellite, one particular godwit flew from Alaska to New Zealand in a single flight, covering 7,500 miles in eleven days. Flying across oceans, through cloud and storm and night, requires a sense of height.

Air pressure is determined by the weight of all the air above you. While air is light, it does have substance, as anyone who has stuck their hands out of a speeding car will confirm. Therefore, the higher you go, the lower the air pressure. At the summit of the Pico de Orizaba, 18,491 feet above sea level, air pressure is roughly half that of sea level. The summit is the highest point in Mexico, and the third-tallest peak in North America, high enough to support a few glaciers within the sunbaked country. At the top of Mount Everest, at 29,029 feet, air pressure is about a third of normal. With such sparse oxygen molecules to breathe in, an average human would pass out in minutes if instantly transported to the top. This is why climbers spend months acclimating to the thin air and make the final push through the "death zone" with supplemental oxygen tanks. Air pressure continues to drop as you fly higher, up to the Kármán line, sixty-two miles above the surface of Earth, when atmosphere gives way to space. Of interest, and known to all scuba divers, water is much heavier than air. To halve air pressure, you have to climb 18,000 feet up. To double atmospheric pressure underwater, you only need to dive thirty-three feet down.

Since air pressure drops with increasing height, you can estimate altitude by feeling the pressure of surrounding air. The paratympanic organ consists of a fluid-filled sac, ringed by hair cells on one side, and connected to the eardrum on the other side. As air pressure changes, the position of the eardrum shifts as well to the detriment of flying babies everywhere. For example, if the surrounding pressure falls, the eardrum will bulge outward, as the trapped air molecules in the middle ear can now push harder with their constant collisions than the less dense air molecules on the outside. This shift in the eardrum deforms the shape of the paratympanic organ, which is then sensed by the embedded hair cells. With the paratympanic organ, birds can maintain a constant flying altitude of +/- ~60 feet, even when flying thousands of feet in the air in the dark.

Semicircular canals appeared in the fossil record between 400 and 500 million years ago. At first, we see just one semicircular canal in ancient creatures. However, subsequent branches of the evolutionary tree show animals with three semicircular canals—including dinosaurs, lungfish, and mammals. Hans Straka, a researcher in Munich, notes that semicircular canals were so advantageous, every animal on Earth has them. Moreover, most have three canals on each side of the head. Exceptions include creatures in the main trunk of the evolutionary tree (the prevertebrates), primitive eel-like mud monsters: the hagfish, which has one semicircular canal; and the lungfish, which has two. He theorizes that the third canal to be added, the horizontal canal, occurred in response to the development of the neck. Prior to that, head movement and body movement were essentially the same, but afterward animals could move the head independent of the body. With the additional degree of mobility, an additional sensor was required.[6]

In a research paper from 2009, Lawrence Witmer and Ryan Ridgely used several genuine *Tyrannosaurus rex* skulls to reconstruct a 3D model of their inner ears.[7] Interestingly, their semicircular canals are recognizable as similar to ours, with three canals that are orthogonal to each other. On the other hand, their cochlea appears as a stubby protrusion from the vestibule, nothing like the elegant spiral that we possess. We can infer that the T. rex had good balance—as you'd expect from a creature shaped like a giant seesaw—but only a rudimentary sense of hearing. For reference, T. rex stomped around Earth during the late Cretaceous Period, becoming extinct sixty-six million years ago, along with 75 percent of life at the time, including the rest of dinosaurs. The terminal event was an asteroid that smashed into the Yucatán Peninsula in Mexico, with an explosive force of one hundred million megatons, which is more than 15,000 times greater than the explosion we'd get from detonating every single nuke currently on the planet.

We humans have two otolith organs—the utricle and the saccule. Fish, amphibians, reptiles, and birds possess a third otolith organ, called the lagena. The lagena seems to function to provide additional vestibular information, and some rudimentary auditory information. Based on ablative experiments, where the lagena was surgically removed in homing pigeons, researchers Le-Qing Wu and J. David Dickson found a third possible function for the lagena: magnetoreception.[8] It's been long known that animals can use Earth's magnetic fields to orient themselves and navigate. The exact mechanism of this remarkable ability is currently unknown and hotly debated. In their experiment, Wu and Dickson first measured nerve activity across different areas of the brain in response to a magnetic field. They found increased neural activity in vestibular nuclei, where the inner ear is first plugged into the brain; the thalamus, which relays sensory information; and the hippocampus, which is important for learning, memory, and navigation. Animals without a lagena did not show those increased responses, indicating that the pigeons required the lagena to be able to sense the magnetic field. In a follow-up study, the research team recorded individual nerves in the vestibular nuclei, in response to changing magnetic fields. The nerve-firing patterns were found to encode the direction, intensity, and polarity of the magnetic fields that were used in the experiment. And those magnetic fields were the same strength as Earth's own magnetic fields. The authors surmised that since these magnetically sensitive cells were located in the vestibular nuclei, the pigeons were combining all the information coming to the nuclei—including acceleration, gravity, and magnetism—to furbish themselves with a biologic GPS system.

While humans cannot sense Earth's magnetism, we do react to very strong magnetic fields. MRI machines use industrial-strength magnets as part of their magical technology to take pictures of the insides of our bodies. The magnet strength inside an MRI machine is about 100,000 times stronger than Earth's magnetic field. There are different strength magnets used in different MRI machines, with 1.5 tesla and 3 tesla magnets being the most commonly used, and 7 tesla machines used more for research. When people are placed inside 7 tesla machines, the inner ear is activated, causing nystagmus and vertigo.

My friend and colleague, Dr. Bryan Ward—together with engineer Dale Roberts and research neurologist David Zee and others—were quite intrigued by that phenomenon, and so they teamed up to explain it. Bryan is currently a professor at Johns Hopkins, in Baltimore. He's naturally curious, and incredibly smart—a winning combination. Bryan told me the story of the discovery:

David Zee spends part of his life in Italy near Siena, and he had been meeting with an otolaryngologist there named Vincenzo Marcelli. And

Marcelli was doing a study where he would put people in a MRI machine, a functional MRI, and he was trying to see if he could look at the areas of the brain that are processing vestibular information. And so he was stimulating the inner ear with a caloric test and using the fMRI to see what parts of the brain might be affected. The advantage he had there was that he was looking at the person's eyes in darkness with infrared video goggles. While looking at the eye movements, he noticed that there was a spontaneous nystagmus that was happening in this 1.5 Tesla MRI machine. He writes in the discussion of this paper that he thought that the magnetic field itself might be stimulating the inner ear. And so he had mentioned this to David Zee. When David Zee returned to Baltimore, he and Dale Roberts began a study of putting healthy humans into the 7 Tesla MRI machine and closely looking at their eye movements with infrared video goggles. And what they found is that all healthy adults that they placed in the MRI machine had a spontaneous nystagmus the entire time that they were in the scanner. And it didn't matter if the scanner was running or not, it just depended on the static magnetic field of the MRI machine. They put people who had no vestibular function in the MRI machine, and those people had no nystagmus and no dizziness. And so that suggested that the effect had to do with an intact vestibular system. Now, there had been some prior studies in rats and mice from Florida State University that suggested similarly that the vestibular system could sense strong magnetic fields from an MRI machine. But this was the first time it had been done in humans. And Dale Roberts designed a series of studies to help sort out the potential mechanism for how this would work. And ultimately decided that the most likely mechanism was something called a Lorentz force. A Lorentz force happens whenever you have an electrical current, and you introduce that electrical current into a magnetic field, it gets a force on the current. So, it turned out that the inner ear is kind of a perfect environment for all this to occur with hair cells, and the resting discharge of the hair cells. And so when you put that electrical current into the magnetic field, it'll get a force in the fluid.

And so that creates a flow of endolymph that then pushes on the cupola of the semicircular canals and generates nystagmus and generates the sensation of dizziness. Interestingly, the sensation of dizziness only lasts for about a minute and a half, and then it goes away. But the nystagmus persists the entire time. The sensation is a weird one, at least personally, what I feel when I go into the scanner is that it's as if I'm lying on the center of a merry-go-round on a children's playground, and so my head is kind of rotating to the left and my feet are going to the right, and the axis of rotation is around my belly button. And that lasts for about 90 seconds or so. And then when I come out of the scanner, as long as I'm in darkness, I feel like I'm going in the opposite direction. Everything reverses. And the nystagmus goes the other way as well. So we think that the primary driver is the electrical current entering the utricle hair cells. The utricle in the human has the highest number of hair cells. And if you get a net electrical current above the utricle, then that'll cause a force in the endolymph right above the utricle and push on the semicircular canals of the superior and the lateral semicircular canal. So even though the current is coming from the utricle, the thing that is sensing the fluid movement is the semicircular canals, and that makes you feel like you're rotating and gives you the nystagmus. We did a study in which we put mice into a powerful MRI machine and the normal mice had the same kind of nystagmus that the humans did. But if the mice didn't have a functioning utricle, there was no nystagmus and presumably no dizziness.

It's a neat story, with a good lesson. Careful observation, coupled with curiosity, a willingness to experiment, and a good grasp of physics, and behold—new discoveries await.

The structures of hearing come onto the scene later than the vestibular apparatus. Fish don't have a dedicated organ of hearing, but they can sense some sounds through their vestibular organs. Keep in mind that the vestibular organs are fundamentally movement sensors, and therefore sound vibrations of sufficient force will activate them. In fact, even human vestibular organs, like the utricle and saccule, are sensitive to loud sounds. We take advantage of that with VEMP testing, which is a type of vestibular test (we'll cover it later) that uses sounds to test the functioning of the vestibular system.

It's believed that a primitive sound-sensing mechanism evolved as animals were making the transition from sea to land, around 380 million years ago.[9] This would have been an early version of the basilar papilla, which is the hearing organ in animals except for most mammals. The basilar papilla varies widely in size and capability across species. In turtles, the basilar papilla is only 1 mm long and can only hear low frequency sounds. Barn owls, on the other hand, have a 12 mm-long basilar papilla, and can hear soft sounds that are inaudible to humans. As nocturnal hunters, they rely on their incredibly acute hearing not only to hear prey, but to locate them. Because two ears are separated in space, sound waves will hit each ear at slightly different times. So, for example, if a mouse is on the right side of the owl, then the pitter-patter of the mouse's paws will reach the right ear before the left. That difference in timing will change, depending on the mouse's location. The owl's brain can use those miniscule timing differences to triangulate the mouse's position, increasing the odds of a successful hunt.

Therians, a grouping that includes placental mammals (and excludes everyone's favorite egg-laying mammal—the duck-billed platypus), are distinguished by having three hearing bones (ossicles), and a coiled cochlea. For comparison, birds, lizards, and amphibians all have one hearing bone. Of interest, there is good evidence that the extra two ossicles that are present in mammals were proto-reptilian jaw bones that were repurposed toward hearing. The close connection between the jaw and hearing is clinically apparent. For instance, many patients who suffer from disorders of their temporomandibular joint have ear pain as their primary symptom, due to close proximity and messy wiring. If you stick your finger in your ear canal, and then open your jaw, you can feel the movement of the jaw joint.

If evolution guided the development of the vestibular system to enable complex, coordinated movements, then are there any vestibular superanimals? A creature with such high vestibular demands that therefore has evolved a superior sense of balance. Paleontologist Camille Grohé was intrigued by this question.[10] To study it, she chose the fastest land animal on Earth. The cheetah is a deadly hunter, capable of sprinting at 60 mph. To catch grazing gazelles at that speed, the cheetah requires sharp vision, rippling muscles, and an extraordinary vestibular system to provide microsecond updates to the brain about the cheetah's position and heading. At 60 mph, at full gallop over uneven terrain, a cheetah without a vestibular system would not be able to see, let alone hunt. To investigate further, Grohé used advanced imaging to create 3D models of the vestibular system of cheetahs and other species of cat. She found that cheetahs had a larger volume of their vestibular than any other of the tested animals, beating out tigers, leopards, lynx, and house cats.

The relationship between vestibular system size and the animal's agility isn't just limited to cats. Fred Spoor, a paleontologist at the University College of London, sought to characterize how vestibular volume correlates with activity across a range of ninety-one different primate species.[11] The primates—including monkeys, lemurs, tarsiers, and apes—were classified on a behavioral activity scale from "fast" to "extra slow." Since it's known that on average, inner ears are larger in larger animals, all analyses accounted for that, by looking at the ratio of vestibular size to body size. Spoor found larger vestibular systems in "fast" primates: those who could effortlessly climb trees, jump between branches, and defy gravity while Tarzaning through the jungle. Spoor points out that sloths, the aptly named lumbering beasts of South America, have relatively tiny vestibular organs.

Spoor isn't just interested in the vestibular systems of living animals. He argues that if we know there is a link between vestibular size and animal activity, then that should hold true for extinct creatures as well. Fortunately, the vestibular system is located deep in the skull, in a protective shell of dense bone. Therefore, it's one of the likeliest parts of any animal to survive over the eons, when the softer bits have long since eroded away. Using this approach, Spoor has been able to look at the vestibular system of extinct lemurs and make reasonable guesses as to their activity levels in the distant past. In addition, Spoor also has a theory that the vestibular system was critical for human evolution.[12] From an anthropologist's perspective, there are two main differences between modern humans and our apelike ancestors: a larger brain, and the ability to walk on two legs. Comparing human, hominid, and ape vestibular systems, Spoor found that modern humans have large vertical semicircular canals, and smaller horizontal canals. While walking seems so natural to us, it's clearly harder to balance on two legs rather than four. In addition, as we walk, our head's natural inclination is to bob up and down in the vertical plane. Spoor argues that the development of large human vertical canals may have been a necessary adaptation for walking. In fact, the other humanlike species that walked— homo erectus (literally "upright man")—has canals shaped like ours, and not like our apelike ancestors. This fundamental connection between walking and the vertical canals may help explain one of the curious features of the vestibular system: the unequal distribution of vertical canals and horizontal canals. We have four vertical canals (two each of anterior and posterior), but only two horizontal canals. If vertical canals are more important during walking, then it makes sense to have more of them.

There are some physical limits on the size of the vestibular system. Since the system is governed by the physical properties of fluid movement, fluid viscosity and the width of the semicircular canal are both important. Recall from physics (or pretend to recall, it's all good!) that per Poiseuille's Law, liquid flow is proportional to the fourth power of the radius of a tube. The upshot is that size really matters because flow varies exponentially. Because of that, the physical size of the semicircular canals is actually fairly consistent across species. While you can literally crawl through the chambers of the whale's heart, their vestibular apparatus is similarly sized to yours.

That also means that below a certain size, the vestibular system simply doesn't work. If the semicircular canals are too small, then resistance to fluid movement is too high, and head movements won't be noticed by the ever-vigilant hair cells.[13] Miniature frogs are the smallest vertebrates on Earth, with the record setter being the New Guinean *Paedophryne amanuensis*, measuring 7.7 millimeters in length, placing it just slightly larger than a housefly. Dr. Richard Essner and colleagues studied pumpkin toadlets, a brightly colored frog that fits comfortably on a human fingernail. The pumpkin toadlet lives under fallen leaves in the Brazilian rainforest, eating tinier invertebrates. Dr. Essner hypothesized that the miniscule frog skulls could not house a functional vestibular system. To test this, they scanned the frogs using microcomputerized tomography and reconstructed the semicircular canals in 3D. They found that frogs do increase the relative size of the inner ear as head size decreases, in an effort to maintain function, but in the smallest frogs, the head is just too small. Pumpkin toadlets don't like to jump, and when the research team studied their agility with high speed cameras, it became clear why. The miniature frogs had terrible balance. Like mannequins shot from a canon, they twisted, flailed, and fluttered, landing awkwardly at best, and 35 percent of the time, landing on their back. The research team was kind enough to make videos of the frog jumps available online.[14] A normal frog jump is graceful, as shown in the first video, with a controlled flight followed by a smooth landing. The miniature frog leaps, conversely, look like a blooper reel. They don't seem to sense how their bodies are rotating in space, resulting in clumsy landings that appear random, including one sad frog who comes down directly on its head. The researchers report that the only way to make a normal-sized frog perform so poorly is to remove its vestibular system.

A quick confession. I have oversimplified things a bit. While the vestibular system is crucial for sensing gravity, we also use joint and muscle stretch receptors to sense the squeeze of gravity. When you are upright, your knees can feel the whole weight of your body (plus that breakfast burrito) bearing down. When you are lying flat, the unburdened knee feels no such force. Therefore, by comparing joint stress in different positions, gravity's vector is easily calculated.

Animals that spend their time underwater or in the air can't take advantage of that ability. For the typical fish, neutrally buoyant, floating in seawater, there isn't appreciable gravitational strain pulling the top part of the body downward. Similarly, in air, without the counter-push of the ground, joints don't get much information to sense gravity's pull. For flying and swimming animals, the vestibular system is critical, as it provides gravitational estimates regardless of environment.

Most flying animals are insects. While birds, bats, and some dinosaurs also evolved the ability to take to the skies, Earth is dominated by insects. There are about one million known species of insect, comprising almost half of all known life. To meet the challenges of flight, executing aerial acrobatics on the order of milliseconds, some insects have come up with clever solutions. Flies must quickly adapt flight patterns to compensate for wind, hungry birds, and swatters. The fly's solution is actually the same solution that planes and spacecraft use to maintain a level trajectory: it uses a biologic gyroscope.[15] Flies have a sensory organ called halteres on each side of their torso, just behind their wings. The halteres are shaped like drumsticks, and they quickly beat up and down with each flap of the wings. They look funny, like knobby, useless, miniature proto-wings. Because they are in constant motion, they have momentum. If the fly is spun out of control, then there is angular acceleration of the fly's body. By sensing deflections in the rhythmic beat of the haltere, a fly is able to sense when it's been knocked off course, responding reflexively with a course correction. Flies seem mundane to us humans: giants who move slowly and think slowly, hearing the flap of their wings as a continuous buzz. In their scaled-down world, air is a viscous fluid that they can push against, soaring, tumbling, and freewheeling through an endless diamond sky. (Author's note: I definitely plagiarized those words . . .). It's only been in the last few decades, with ultra-high-speed cameras, that we have been able to really glimpse into their world.

It's a beautiful thing to contemplate. From a primordial need to sense gravity, we inhabitants of planet Earth have evolved the capacity to balance ourselves, to steady our vision, and to hear the wonderful world around us.

PART 2

HOW IT ALL WORKS

3

A Head Full of Hair Cells

I n this chapter, we will explore the inner workings of the remarkable
organs that bestow the gift of balance. We are going to learn about the
hair cell, the basic sensor that detects movement, and how our bodies
manage to use that one building block to allow for a variety of needs, including enabling hearing and balance. In so doing, we'll find out why the auditory
and vestibular apparatuses are located in the same space. We'll appreciate
the fine structure of the inner ear, which reflects eons of adaptation that
formed an intricate "labyrinth" within our head. And finally, we'll cover why
we actually do have "crystals" located in the inner ear.

As we saw in the chapter 2, the hair cell is the technological breakthrough
that powers the auditory and vestibular systems. From its start as a simple
gravity sensor, the hair cell became specialized to allow for a broad array
of functions. It's a great example of the inherent ingenuity and efficiency of
nature, where a new use is found for biologic machinery. Turns out the ability to sense microscopic movement is rather useful!

To see how the hair cell works in so many different capacities, let's explore
the structure of the inner ear. There are three sections. Centrally is the
vestibule. On the front side, extending outward is the spiral of the cochlea.
The cochlea is rather beautiful, a striking spiral of two-and-a-half turns.
Off the back of the vestibule are the semicircular canals. They amaze as
well, three perfect arcs, each at a ninety-degree angle to the other two.

Surrounding this assemblage of shapes is a dense capsule of protective bone. The spirals and circles are not just for show, as form follows function, and these shapes allow for very specialized functions.

YOU MIGHT BE HEARING THINGS: THE COCHLEA

We will first consider the cochlea, the organ of hearing. Sound waves are ubiquitous, traveling through the air as energy is released by object collisions or vibrations. This could be a plate crashing to the floor, a guitar string being plucked, or vocal cords quivering during a high note. The eardrum functions like the diaphragm of a microphone, vibrating in tune with colliding sound waves. The name "eardrum" is no coincidence, similar to the way a drum consists of a drumhead stretched taut over a frame, the eardrum is held in place along its outer edge. As the eardrum moves, those vibrations are transmitted along the three hearing bones to the cochlea. Mechanically, there is an advantage to that setup, which is why people with holes in their eardrums, or absent hearing bones, experience hearing loss. By collecting sounds from a relatively large area (the eardrum) and concentrating them into a small area (the bottom of the stapes bone), the energy is amplified. Furthermore, the shape of the first two hearing bones—with the malleus being longer than the incus—results in energy concentration as well, due to the principles of levers. This force concentration is necessary, because the sound vibrations need to transition from the air-filled space of the middle ear to the fluid-filled space of the cochlea. If you've ever tried to shout at someone underwater, you know that most sound energy is deflected at the surface, which is why they don't hear you. This makes sense, sound waves depend on vibrating molecules, and air is a lot less dense than water, which means that there are far fewer molecules in a given area. It takes more energy to vibrate the densely packed molecules in fluid than the sparse air molecules. We need a way to enhance the power of sound waves before they enter the cochlea, and that is the primary function of the middle ear.

The stapes bone sits atop an oval-shaped membrane and bounces up and down with sounds. Those vibrations then propagate along the length of the cochlea, like a ripple spreading across the surface of a pond. Inside the spiral of the cochlea runs a partition known as the basilar membrane. Georg Von Békésy, a Hungarian scientist, won the Nobel Prize in 1961 for his discoveries about the natural function of the cochlea. One of the fundamental properties of sound is pitch, which correlates with the wavelength of the sound wave. By dissecting the cochlea, and covering the basilar membrane with visible silver flakes, he was able to observe that different pitches of sound vibrated the cochlea in different regions. Just like a piano, the low frequency sounds come from one end of the cochlea, and the high frequency sounds from the other end. A prism diffracts light into different wavelengths, seen as the colors of the rainbow. Békésy discovered that the cochlea is an acoustic prism, decoding complex sounds into their constitutive frequencies. This is possible because the basilar membrane is stiff and narrow at the base—where high-pitched sounds are heard—and wide and flexible at the apex—where low-frequency sounds are heard. Each region along the spiral cochlea vibrates in tune with a specific pitch. That feature is called tonotopic organization, and it is the foundation of the most revolutionary hearing-restorative technology of all time: the cochlear implant.

At first, the idea of threading an array of electrodes into the delicate interior of the cochlea was considered sacrilegious by ear surgeons. However, nerves communicate with each other by electric impulses, which raised the possibility that an electric current could restore the sensation of sound to those with a faulty cochlea. In 1957, a pioneering French team—André Djourno and Charles Eyriès—implanted a deafened patient with a metal wire against the hearing nerve. Through an induction coil, impulses were sent through the wire to the hearing nerve. Remarkably, when current was applied, the patient heard a sound that resembled a squeaky wheel. Reports of this achievement were brought to Dr. William House, a trailblazing ear surgeon in Los Angeles, by one of his patients (goes to show that you should ALWAYS listen to your patients). Dr. House's curiosity was sparked, and he is credited with performing the world's first cochlear implant in 1961. Many years of research followed, and Dr. House's original design, with just a single electrode, was refined into multichannel devices. From their first FDA approval in 1984, cochlear implants have become the most successful neural prosthesis to date, completely redefining what is possible in the treatment of hearing loss. And they work because of the tonotopic organization of the cochlea. Since multiple electrodes are spaced out along the interior passage of the cochlea, different electrodes in different locations can be activated for different sounds. So, the minicomputer inside the cochlear implant translates complex sounds into patterns of activation for the various electrodes. This results in a sound that isn't quite perfect but generally allows a cochlear implant recipient to hold a face-to-face conversation.

Now, let's get back to the normal functioning of the cochlea. As we left things, sound vibrations were causing the basilar membrane to vibrate, like a plucked guitar string. These vibrations can be felt by hair cells. But it turns out that the story is not quite that simple. That's because there are two types of hair cells inside the cochlea: inner hair cells and outer hair cells. The hair cells are arranged in neat rows, one row for the inner hair cells, and three rows for the outer hair cells. Their spacing is regular, like soldiers standing at attention. Now, the inner hair cells are connected to the hearing nerve and send impulses along the nerve when their section of the basilar membrane vibrates. So, what then is the function of the outer hair cells, who far outnumber the inner hair cells?

To understand why we have outer hair cells, it's useful to put the size of the transmitted sound vibrations into perspective. In the seminal textbook, *Schuknect's Pathology of the Ear*, Liberman, Rosowski, and Lewis explain—and note that I am paraphrasing for clarity—"The amplitudes of motion are extremely small: at very loud sounds, stapes motion is roughly 100 nanometers, almost too small to see with the light microscope. At the threshold of hearing, stapes motion is 10,000 times less, or 0.001 nanometer, less than the size of a hydrogen atom."[1] How are hair cells able to sense such infinitesimally small movements?

As a child, I petitioned my parents for a trampoline. While my parents never caved into purchasing a "child maiming lawsuit machine," lady luck was kind, and a neighbor did. Now, as any trampoliner can attest, there's a secret to jumping as high as possible. You must first get into a rhythm of jumping. Next, jumps are timed such that you are hitting the surface of the trampoline when it's already been depressed, so that your weight pushes it down even further, and the enhanced recoil launches you into the air. If the timing is off, then your feet hit the trampoline as it's coming up, and the resultant force collision usually results in a quick buckling of the knees, and a fall. But, if you really want to soar high, then you need help. And that is best provided by a friend, using their weight to push the trampoline down right before you jump. And with multiple friends, it's quite possible to launch a small child clean out of the neighborhood. In the cochlea, the outer hair cells are those friends, enhancing the natural vibrations of the basilar membrane. They do this thanks to a remarkable protein called prestin, which lines the cell walls, and changes their shape to shorten and lengthen the cell, like a spring. This extra oomph improves inner hair cell ability to hear soft sounds, and to discriminate between different pitches of sound. It turns out that the sound of thousands of inner hair cells jumping up and down is audible, provided you have a sufficiently sensitive microphone. Measurement of this sound is called an "otoacoustic emission," and it's the basis for many newborn hearing screenings in the United States.

To this point, we've explained that hair cells can feel vibrations. But how do they actually accomplish that feat? The answer lies in the hairs, which come out of the top surface of the cell, like bristles on a brush. They aren't actually hairs—just microscopic filaments called stereocilia. When the fluid around them moves, the stereocilia sway like reeds in a river. And that swaying starts a cascade of events, like a relay race, that results in the hair cell sending an activation signal to its connected nerve cell. Now, I've tried to come up with the best way to explain this process, and here's what I came up with: a toilet. It's like a toilet flushing. Bear with me. When you press the flush handle, you are tugging on a chain that pulls up the flapper (or flush valve) located under all the water in the tank. When it opens, the water that it has been holding back rushes through the valve into the toilet bowl. The stereocilia also have valves, called ion channels, along their length. And similar to the flapper in a toilet, these valves have a chain attached to them, which can pull them open. These chains are called tip links, and they are connected to adjacent stereocilia. That means that if the stereocilia bend enough in one direction, they will spread apart, and tip links will pull open the valves. When the valve opens, positively charged ions will rush into the cell, changing the electric potential of the cell from negative to positive (called depolarization). The ions rush in for the same reason that water rushes into the toilet—energy has already been spent storing energy, and all that potential is just waiting for an opportunity to dissipate. In the toilet's case, the energy is gravitational. With hair cells, the energy was spent building up an ion gradient, so that the positively charged ions were concentrated outside the cell, and the negative ones inside the cell. This pent-up energy can't last, and as soon as the gates open, the positively charged ions—potassium and calcium—come rushing in. If enough potassium and calcium enter the cell, then the cell will send a signal—in the form of glutamate—to its connected nerve cells. Each of the ~3,500 inner hair cells are connected to about ten nerve cells. If the hair cell signal passes a certain threshold, then the neuron will send the message along the brain that a certain sound is being heard.

In his 1928 novel *Point Counter Point*, Aldous Huxley provides a neat summary: "The shaking air rattled Lord Edward's membrane tympani; the interlocked malleus, incus, and stirrup bones were set in motion so as to agitate the membrane of the oval window and raise an infinitesimal storm in the fluid of the labyrinth. The hairy endings of the auditory nerve shuddered like weeds in a rough sea; a vast number of obscure miracles were performed in the brain, and Lord Edwards ecstatically whispered 'Bach!'"[2]

Now that we have seen how hair cells are the secret ingredient in the hearing apparatus, let's look at what they do in the vestibular system. We will start with the semicircular canals.

A THREE RING CIRCUS: THE SEMICIRCULAR CANALS

The semicircular canals are quite unique. Six little circles, at right angles to each other, located deep inside the skull. They are connected, as part of each circle includes the vestibule, the central chamber of the inner ear. And two of the canals, the posterior and superior (each canal is named simply for its location), share part of their arc. Nature, it seems, loves design efficiency. Why do we have these elegant, interlocking rings? A recurrent theme in biology is that "form follows function," and it's certainly true here. In order to sense head turns, all vertebrates have at least one semicircular canal, and tetrapods (amphibians, reptiles, birds, and mammals) have three. We previously saw how hair cells, acting as microscopic movement sensors, enabled hearing. Let's see what they do for the vestibular system.

In the semicircular canals, hair cells are embedded in a saddle-shaped structure, called the crista. The stereocilia (the microscopic hairs of the hair cell) protrude upward from the crista into a flexible gelatinous wall, called the cupula. Each semicircular canal has one region that houses the crista and cupula: a dilated bulb called the ampulla. In the horizontal and superior canals, the ampulla is at the front part of the semicircle, in the posterior canal it's on the bottom. Just like in the cochlea, this sensory part of the inner ear that contains hair cells is bathed in a fluid called endolymph. There's a membrane that envelopes all the inner ear sensors and separates them from the surrounding perilymph fluid.

Now that we've got the jargon out of the way, let's see how it works. We'll use the right horizontal canal as an example. The horizontal canal, as the name implies, is in the horizontal plane, just like the Karate Kid's headband. When you turn your head to the right, everything fixed to the skull, including the horizontal canal and its cupula, turn as well. But the endolymph inside the canal is a fluid, and it has inertia. That means that it wants to

Cupula

Ampulla

Crista

Hair Cell

Stereocilia

Vestibular Nerve

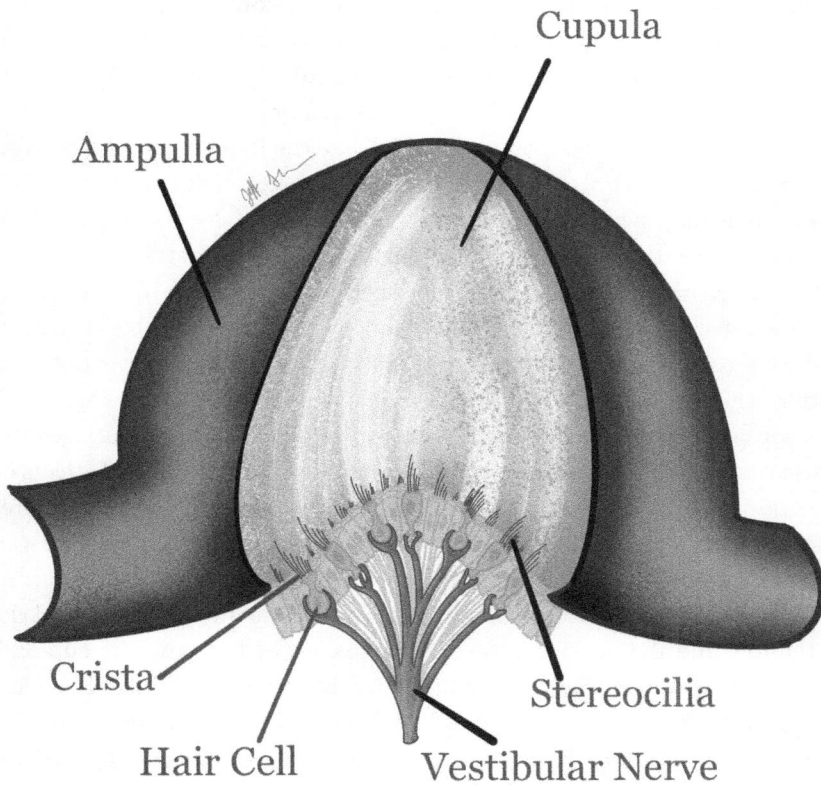

FIGURE 3.1 The ampulla of the semicircular canals.

stay still. Picture if you had a water bottle, and you held it at arm's length, on its side, with the opening facing left, and spun around to the right. Well, I've done that experiment, and guess what? The water spurts right out of the bottle onto the floor—because inertia is trying to hold it in place, where the bottle was. But the inner ear doesn't have an opening anywhere, so thankfully we don't leak endolymph when we turn. Therefore, let's give the bottle a cupula, which will be a cover of plastic wrap that we stretch over the opening, and rubber band to hold it in place. Let's assume our contraption is watertight. We'll repeat our experiment, holding the bottle sideways, with the plastic wrap on the opening on the left of the bottle and turning our body to the right. The water inside still has inertia. When we turn, the plastic wrap will be pushed outward by the water, and we will be able to see it bulge. Similarly, when the head is turned to the right, the cupula will bulge outward, due to the inertial force of the endolymph pushing against it. And if the head was turned to the left, it would bulge inward. Recall that the hair cell stereocilia stick up into the cupula, like toothpicks stuck in Jell-O. So, each time the cupula bulges inward or outward, the stereocilia are pushed or pulled together with it. Just like the hair cells in the cochlea, vestibular hair cells also have channels along their sides that get pulled open when they sway, causing an electric current to flow into the cell. As detailed above, this sparks a chain reaction that culminates in the signal being passed along to a dedicated nerve cell, and then onward to the brain.

Now, have you ever experienced a night clinging to a bathtub as the world violently rotates around you, questioning why you downed a bottle of bottom-shelf tequila from a plastic bottle? Was that night so bad that to this day, you still can't stomach the taste of tequila, even when camouflaged in a margarita? Your cupula is likely to blame. Alcohol has a number of effects on the brain, like disinhibition, clumsiness, and stumbliness. However, there's also evidence that alcohol directly diffused into the cupula causes the vertigo of intoxication.[3] Keep in mind that alcohol is less dense than water. So, if you were to pour pure alcohol and water into a cup, the alcohol would rise to the top, and the heavier water would sink to the bottom. The cupula is normally the same density as the surrounding endolymph. But, when alcohol gets into the cupula, it becomes light and tries to float away. Depending on your body position, this will trick your semicircular canals into thinking that your head is spinning when it isn't, causing your eyes to respond (nystagmus), and giving you the spins (vertigo). Unfortunately, this is the main way that most American college kids learn about their cupulas . . .

Each semicircular canal has an estimated 7,600 hair cells, all facing the same direction.[4] If you turn your head slowly, only a few will get activated, and you'll get a weak nerve signal. If you whip your head around with force, many more will get activated, and the nerve signal will be stronger. In that way, the strength of the turn (the rotational acceleration) is encoded into the nerve signal. The system is quite sensitive. Doctors Lloyd B. Minor, Timothy E. Hullar, and David S. Zee report that humans can sense acceleration as low as 0.1 degrees per second squared. As they report, if you were to completely swivel around in a chair at that acceleration, it would take ninety seconds.[5]

How does the vestibular system achieve such remarkable sensitivity to faint movement? Researchers Charles Oman, Edward Marcus, and Ian Curthoys used a mathematical model to calculate the minimum distance that the cupula needs to be moved from its resting position that can be sensed by hair cells.[6] They estimated that value at 520 angstroms. An angstrom, like a mile or a centimeter, is a unit of length. One angstrom is an incredibly small distance—it's defined as one ten-billionth of a meter. For reference, the diameter of a hydrogen atom— the smallest element—is one angstrom. Strands of human hair (not hair cells but actual hair, sprouted from the scalp) are approximately 200,000 to 400,000 angstroms wide. So, the cupula only needs to move about a quarter of one percent of the width of a human hair for a hair cell to notice that movement and initiate a nerve signal. In the same paper, Oman, Marcus, and Curthoys calculated that on the other end of the spectrum, the cupula would still be sensitive to much larger cupular movements, up to ten micrometers (100,000 angstrom).

Each of the six semicircular canals has a partner in the same plane, on the opposite side of the head. So, the right horizontal canal is in the same plane as the left horizontal canal, and the right superior canal is in the same plane as the left posterior canal. Since each canal is the mirror image of its partner, that arrangement ensures that no matter which way the head moves, there will be a semicircular canal that will get maximally stimulated. Of course, most movements are not perfectly in the plane of a specific canal and will therefore result in a partial stimulation for several canals, which is then translated by the brain into the correct vector of movement.

There's another interesting piece to this puzzle. Neurons communicate in an all or nothing event, called an action potential (aka spike aka impulse). So, the nerve bundles coming out of each semicircular canal—called the ampullary nerve—have a limited vocabulary for transmitting messages. They can either send an action potential, or not. But each canal is capable of telling the brain much more information. The horizontal canal can tell when you are turning to the right *or* to the left, and also the strength of the turn. How does the ampullary nerve, with only one word in its vocabulary, achieve this? The answer lies in the timing. Each nerve impulse is incredibly quick, lasting only one or two milliseconds (1,000 milliseconds = one second). That's about the time it takes a fired bullet to exit the barrel of a gun. Nerves do need to rest quickly after each impulse, which is called a refractory period. Two observations follow: the slowest that these nerves can fire is zero impulses per second, and the fastest that they can fire is about four hundred impulses per second. One core principle of vestibular physiology is that the nerves have a resting firing rate. They are constantly sending off impulses—about eighty to one hundred per second—even when your head isn't moving. This strategy burns through energy, but it's worth it, as it allows the nerve to convey all the necessary information to the brain. Let's go back to our prior example, the right horizontal canal. When you turn your head to the right, the hair cells get bent as the cupula is pushed forward, toward the vestibule. This direction of movement is excitatory, because it results in the ion channels being pulled wide open (For the curious, the stereocilia are oriented relative to a special, giant stereocilia called the kinocilium. Due to the setup of the tip links, when the stereocilia are bent towards the kinocilium, the channels get pulled open). Therefore, the hair cells will pass this message along to the nerve cell (by releasing a messenger chemical called glutamate), and the firing rate will increase. When you turn your head to the left, then the cupula gets pulled away from the vestibule, like a sail in the wind. The stereocilia are deflected in a way that makes it less likely for the ion channels to open. This results in fewer nerve impulses being sent. And no matter which direction the nerve fires, the change in firing is proportional to the head movement.

Now, you may have noticed an asymmetry here. The nerve firing rate can increase by ~300 impulses per second (from ~100 to ~400) but can only decrease by ~100. That's important, because it means that, from the perspective of each canal, the two directions of movement are not equal. A movement in the activating direction results in a much stronger change in nerve signal than a movement in the inhibiting direction. For our favorite canal (I'm being facetious here, I love all the canals equally, like children)— the right horizontal canal—that means that a rightward head turn is excitatory (stronger signal), and a leftward head turn is inhibitory (lesser signal). There are a few really important implications for that. First, even though the canals exist in pairs, for each head movement there is going to be one canal that sends most of the signal to the brain. So, both horizontal canals are affected by horizontal turns (shaking your head "no"), but not equally. Second, if you lose function of the inner ear on one side of your head, you will be missing a lot of information, but not all information. That's because even with head turns to the bad side, you'll still get the inhibitory information from the good side. So, one inner ear will still tell you about any type of head movement, but it will be better at sensing head turns toward its side than those away from its side. We'll explore the clinical implications of that in later chapters. Finally, this asymmetry means that we can test each canal separately. So, even though any horizontal movement will result in changed nerve activity in both horizontal ampullary nerves, since most of the response comes from the side being activated, we can effectively test each side separately.

Until now, we've used the horizontal canal to explain things. The vertical canals—posterior and superior—roughly work the same way. But, both are offset from the center of your head, by about forty-five degrees. So, if you look straight up at the sky, you are activating both of your posterior canals equally (and inhibiting both superior canals). So, while we talk about six semicircular canals, they are not evenly divided: four of them are vertical, and only two are horizontal. Why is that? One theory is that we want more information in the vertical plane because there are naturally more head movements that are up and down, as compared to side to side. This has been confirmed experimentally, with research subjects wearing sensors while carrying out regular activities, like walking and running in the woods.[7]

Just like the horizontal canal, each of the vertical canals has a specific head movement that results in maximal activation. To activate the right superior canal, you would need to turn your head to the left by forty-five degrees, and then quickly tilt your head downward. In that position, the upward head movement would activate your left posterior canal. Similarly, to activate the left superior canal, you would turn your head to the right by forty-five degrees and then tilt it down quickly. Again, the upward head movement in that position would activate the right posterior canal.

Let's dive into one more nuance. We already said that the horizontal canal gets activated with head turns toward that side, in the horizontal plane. (That head movement is called yaw, as our heads use the same naming convention as aircraft do. Similarly, pitch is an up and down movement—like nodding yes, and roll is a tilting movement—like the head moving toward the shoulders. And we reviewed earlier, in excruciating detail, how that occurred because the cupula was pushed into the vestibule. But, for each of the vertical canals, the movements that I described as excitatory resulted from the cupula being pushed away from the vestibule, not toward it. How can that be? J. Richard Ewald noticed this during his pigeon experiments, and realized that for the vertical canals, the cupula being pushed away from the vestibule is excitatory. This is because hair cells are oriented differently in the vertical and horizontal canals. It's still the case that the stereocilia movement toward the unique kinocilium is excitatory, it's just that the position of the kinocilium is different for the vertical canals. Interestingly, this arrangement creates some simplicity: every head movement that is in the plane of a semicircular canal, and toward the side of the head of the canal being used, is excitatory.

Let's review. We took a deep dive into the microscopic functioning of the vestibular hair cells inside the semicircular canals. In so doing, we got to appreciate how that incredible biologic machine senses rotational head movements. We also saw how the body can arrange its cells into wondrous forms, to achieve specific and highly demanding tasks. We'll come back to the design of the semicircular canals in future chapters, because the diseases that we'll discuss result from specific malfunctions of this system. But first, we will explore the otolith organs—the utricle and saccule—and how they use hair cells in a unique way to provide additional vestibular information.

WHAT'S UP, DOC? THE OTOLITH ORGANS

Among the canon of classic science fiction films, *Fantastic Voyage* will always hold a special place for me. Released in 1966, it promised: "this film will take you where no one has ever been before; no eyewitness has actually seen what you are about to see." This hitherto unknown place was not outer space, but instead "inner space," which we learn is the inside of the human body. In order to save the life of an injured man who holds key information that may turn the tide of the Cold War, a team consisting of physicians, military personnel, and Raquel Welch is assembled. The team boards a high-tech submarine called the Proteus and are promptly shrunk to microscopic size. Once inside, they have little time to marvel at their unique perspective of the human body, as a series of calamities befall the intrepid crew. In this desperate race against time, the crew decides to take a risky shortcut: across the inner ear. Cut to a dramatic scene of the submersible navigating through a cavernous, curved tunnel, past thousands of luminescent cells, bathed in shades of deep greens and blues. However, we learn that the inner ear is incredibly dangerous for our microbe-sized protagonists, as any ambient sound will be experienced by them like an earthquake. Unfortunately, "reticular" matter from a prior dash across a lymph node has clogged the engine intake valve, forcing a pit stop to repair the damage. Using scuba gear, the team exits the craft and clears the engine. Just then, as luck would have it, someone in the operating theater drops a metal instrument on the floor. The clanging causes massive tremors inside the ear, throwing poor Raquel into the tentacles of a nearby cell. Her crewmates manage to extricate her, just before a swarm of marauding antibodies overwhelm them.

What would it look like if we could shrink ourselves into the microscopic realm, and take a tour of the inner ear? Let's land our magic school bus in the vestibule—the central chamber of the inner ear. Looking up, we would

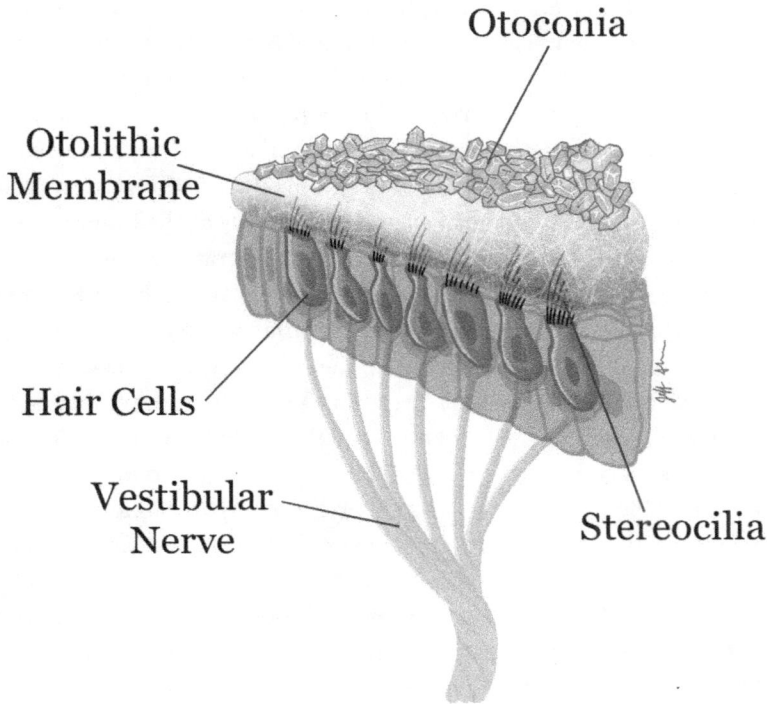

FIGURE 3.2 The otolith organs.

see the undersurface of the stapes footplate, like the hull of a massive ship. With sound, this hull bobs up and down, sending reverberations to the cochlea, whose opening is visible in the distance. Looking around, we'd see the mouths of five other tunnels, leading to the semicircular canals. Recall that two of the canals share a part of their arc, so the three canals result in five openings into the vestibule, not six. But perhaps the most striking thing we would see would be two structures covered in a forest of crystals. These are the otolith organs: the utricle and the saccule. The utricle is suspended horizontally, like a platform, and the saccule is vertical, covering a concave recess. Both are bound by membranes that partition the sensory organs of the inner ear from the surrounding fluid. As we now know, if our bodies are pouring energy, space, and resources toward a highly specialized structure, they must be getting a return on their investment.

Inner ear crystals—otoconia—are present to solve an engineering problem. The entire apparatus of the semicircular canals—the cupula, the crista, and the membranes—all have roughly the same density as the surrounding fluid. That means that at rest, without movement, they will float in place, and there will be no hair cell activation. So, they can't sense gravity or the head position relative to gravity. Furthermore, due to their geometry, they are most sensitive to rotational movements, and not to straight movements. Therefore, crystals, which are about 2.7 times denser than the surrounding fluid, are needed to extend the sensory capacity of the inner ear.

Both the utricle and the saccule have similar arrangements. The forest of crystals lays atop a jelly layer, which in turn is atop a bed of hair cells. Gravity exerts a constant force on the otoconia, pulling them down like a stone in a pond. When the head is tilted, the crystal layer will slide en masse, putting tension on the gelatinous layer. This will pull on the embedded hair cell stereocilia, sparking the cascade of cellular events that results in a nerve signal being sent out. This sliding of the crystals will also occur for linear accelerations. Picture you are in a cafeteria, holding a plastic tray. And for lunch today, rather than your typical soup and salad, you decide to buy a loose assortment of candy, which you place in the middle of the tray. If you hold the tray steady, the candy remains still. But picture what would happen if the tray tilted, or if you suddenly lurched forward (a linear acceleration). Your precious candy lunch would follow the tilt or the lurch—just like the crystals do. And of course, each time the candy gets too close to the edge of the tray, you shout "noooo." The otolith organs function similarly, constantly providing their nerves with information regarding tilts and movements.

There is another difference between the semicircular canals and the otolith organs. Unlike the semicircular canals, the hair cells of the utricle and saccule are not all oriented in the same direction. Instead, their polarity is determined by a dividing line that traverses the center of each organ. That means that the utricle or saccule on each side of the head provides information about tilts or movements in any direction. They are not as biased as the canals in providing more information about movements toward their side of the head, rather than away from it.

There are an estimated 33,000 hair cells in the utricle, and 18,000 in the saccule.[8] They both sense acceleration. However, it's not the rotational acceleration that the semicircular canals sense (like on a merry-go-round). Instead, they sense linear accelerations, which means speeding up in a straight line trajectory, like in an elevator. Due to their positions, the saccule is better at sensing vertical accelerations, and with that the most dominant vertical force of all: gravity. The utricle is better-suited to sense horizontal acceleration, as occurs when you slam the gas pedal in a car (unless, of course, that car is my former college-mobile, a sky-blue Toyota Previa minivan from the 1990s. Some accelerations are too faint for even the utricle to sense).

Our eyes do respond to these linear accelerations. If you are suddenly jolted two feet to your left (without turning at all), your eyes will respond with a countermovement to the right, to keep vision steady (no net movement). Of course, our whole body can move suddenly to the right or left, but our eyes can also rotate. So, the eye's response to either a rotation or a linear movement has to be a rotation. Since this reflex originates with vestibular sensation, and ends with eye movement, it's also called the vestibulo-ocular reflex. And because linear movements are also called translations, it's called the translational vestibulo-ocular reflex. This is similar to the angular vestibulo-ocular reflex (which we will explore further in chapter 4), generated by the semicircular canals. Now, the angular vestibulo-ocular reflex is far more important for everyday functions, and almost all research has been focused on it, so it's the default vestibulo-ocular reflex. Because of that, it's usually not called by its full name—angular vestibulo-ocular reflex—instead, it's just referred to as the "VOR."

For reasons that are not well understood, the translational VOR is weaker than the angular VOR. For the angular VOR, for each degree of head turn, the eyes rotate the same amount. So, a five-degree head movement to the right is counteracted with a five-degree eye turn to the left. Mathematically, that means that the gain of angular VOR is one (it's just the ratio of eye to head movement). For the translational VOR, the gain is weaker. One study estimated it at ~40 percent of the ideal eye movement.[9] There are complexities here, because gains are dependent on the speed of the turn/jolt (called frequency), and other factors like target viewing distance (how far you are from the thing you are looking at). Even though both angular and translational VOR change a bit with viewing distance, it makes sense for the translational VOR to have a wider range. Consider that you are looking at two targets, one five feet away, and one five hundred feet away. Our homework is to calculate the ideal VOR gain for the two different head movements (bear with me. . . . Or skip the math. I'm not here to tell you what to do). For the angular VOR test, we'll rotate the head five degrees to the right. In both cases, at five feet and at five hundred feet, the VOR should tell the eyes to rotate about five degrees to the left. Now, as the target is further away, more of the visual surround will fit onto the fovea (the part of the retina, in the back of the eye, where most seeing is actually done). So, you can get away with a slightly lower gain for faraway objects. Contrast that with the demands of the translational VOR. Instead of a quick head turn to the right, we are going to laterally move the whole body to the right by 3.5 inches. I've chosen that amount, because if my hazy recollection of seventh-grade trigonometry is correct, at that distance you'd have to rotate your eyes five degrees to compensate and keep staring at the target. But at five hundred feet, for the same 3.5 inch translation to the right, your eyes should rotate less than a tenth of a degree. Our geometric analysis shows that we should expect a higher gain from the angular vestibulo-ocular reflex.

The utricle also senses tilts. If you touch your ear to your shoulder, technically that movement is called a "roll," and it activates the utricle. During a roll, the tangle of ear crystals slides atop the carpet of hair cells, with more displacement seen with larger tilt angles. The utricle responds by directing eyes to "counter-roll," which basically means that they tilt in the opposite direction. To add some complexity: while rolling your head to your shoulder you actually activate both your semicircular canals (because it's a head rotation) and your utricle (because the head isn't level anymore, so the crystals start sliding). However, the semicircular canal response settles down quickly, so if you maintain your head in the tilted position (called a static tilt), that's really a utricle-only response. Dr. Amir Kheradmand is a neurologist and vestibular researcher at Johns Hopkins. He studies the utricle by measuring eye movements in response to a static tilt (called the ocular counter-roll). He has calculated that the gain of the counter-roll is about 15 percent, far less than both the angular and translational VOR. So, if I tilt my head to my right shoulder by thirty degrees, my eyes will twist to the left about five degrees to compensate. The response is present when your vestibular system works well, and absent when the vestibular system is broken. Currently, we don't routinely test the counter-roll in the clinic, but Dr. Kheradmand has developed an automated system for measuring eye roll. Therefore, in the future we might routinely be able to understand someone's utricle by just tilting their head!

Now, just to wade back into the weeds for a second, why is the gain of the ocular counter-roll so low? Fundamentally, doesn't it make sense for the brain to keep things simple, with an opposite eye movement to every head movement, so that there is no net eye movement? There are a few possibilities as to why the counter-roll might not need to function that way. First, the counter-roll isn't the utricle's only response. There is also a compensatory head tilt, and a compensatory up/down eye movement (the lower eye tilts up, the higher eye tilts down, which helps level the visual world). So, the counter-roll has helpers, and doesn't need to do everything itself. Another reason is that unlike with head pitches and yaws, with a head roll, you aren't really moving the fovea of the eye away from the visual target, you are just rotating the image on the fovea. And the fovea can still make sense of a rotated image (like holding a photograph at an angle). Taken together, the gain of the counter-roll isn't one because it doesn't need to be.

Before moving on, we are going to explore one more utricular mystery. In 1907, Albert Einstein described the "equivalence principle." The principle states that acceleration due to a gravitational field is the same as acceleration produced by any other force. For example, on Earth, we all experience gravity as a force pulling us toward the center of the planet at 9.8 meters per second squared. At that gravity level, an apple falling from a ten-foot tree would take 8/10ths of a second to fall and hit the ground traveling at 7.7 meters/second. (Author's note: you may be wondering why I am using a mash-up of imperial and metric systems of measurement, and the answer is that I want everyone around the whole world to feel comfortable (or uncomfortable?) reading this book). According to the equivalence principle, if you were on a rocket ship accelerating at a constant rate of 9.8 meters per second squared, and this was a fancy rocket with a ten-foot tall apple tree, then falling apples would act exactly the same as they would on the surface of Earth. They would take the same 8/10ths of a second to fall. According to the equivalence principle, the utricle should not be able to tell the difference between a roll (head moving toward the shoulder) and a translation (head moving to one side without turning). That's because both exert the same force, displacing the otoconia crystals the exact same amount, deflecting the hair cells by the same degree. This results in ambiguity. How does the brain sort out if a specific movement is a tilt to the right or a translation to the left, when from the perspective of utricular hair cells, they are the same?

This was a complete mystery for a long time, but researcher Dora Angelaki may have figured it out.[10] She examined the eye responses of rhesus monkeys to both tilts and translations under normal conditions. As expected, during a tilt, the eyes counter-rolled, and during a translation, there was a horizontal, compensatory movement of the eyes. She then repeated the tests after surgically plugging up the monkey's semicircular canals, rendering them nonfunctional. When those monkeys were tested, Angelaki found the same responses for translations. However, tilts didn't cause the eyes to counter-roll. Instead, the eyes responded to the tilt as though it were a translation, with a horizontal correction. She concluded that semicircular canal information is the key. During a tilt, both the utricle and the canals are activated, whereas during a translation, only the utricle is activated. Therefore, the brain just has to check if there has been any canal activity whenever the utricle sends a signal. In the experiment, without the companion signal from the semicircular canals, tilts were mistaken for translations.

Now that we completely understand the otolith organs, let's explore an example. Researcher Martha Bagnall has shown that in zebra fish, otoconia aren't present just to make the inner ear glitter, like polished quartz. They are necessary for survival. Dramatic pause. Dramatic pause. Dramatic pause. Next. (Author's note: growing up, my parents bought me a toy microscope. It was shiny red, and it came with a few prepared slides of plant cells. Years later, I of course realized that my parents were trying to turn me into King Nerd, Lord of Insecurity. And fortunately for them, with my coke-bottle glasses and the metal cages affixed to my teeth, they didn't have much work to do).

Light microscopy is foundational, but there's an inherent problem. To see something with light, it has to be transparent enough for light to pass through. Microscopes are only able to see a thin slice at a time. Cross sections are useful but are limited by what is included in each picture. However, with modern techniques, it's possible to take an object, cut into hundreds or thousands of razor-thin slices with a microtome (a special knife), digitally capture each slice, segment each part of the anatomy, and then reconstruct a 3D model of the object. It's time consuming work, but the payoff is well worth it. Using this technique (and borrowing a dataset from a colleague), Bagnall was able to rebuild the circuitry of the zebra fish vestibular apparatus, from utricle to brain. She mapped out each hair cell, their connecting nerves, and the brain neurons connected to the incoming vestibular fibers. Furthermore, because stereocilia (the hairs) on each hair cell are visible under electron microscopy, Bagnall was able to infer function from form, and knew which direction of movement would stimulate each hair cell. The research team then reconstructed a map of the utricle and the neurons.

When scared, little fish like the zebra fish escape. They quickly scurry away from any perceived threat. Central to this escape behavioral reflex is the Mauthner cell. The activation of the Mauthner cell is the trigger for the escape reflex. It's like the eject button for a fighter pilot. Once pushed, a cascade of events is initiated, and there is no going back. With her digital readout of a zebra fish brain, Bagnall found that the utricular neurons connect to the Mauthner cell. Based on this, she conducted an experiment. Two types of zebra fish were tested for their escape reflex: normal zebra fish and mutant fish without functional otoliths. Each fish was placed in a small dish of water, and then quickly moved to one side. The normal zebra fish reacts appropriately and gets out of there as quickly as possible. A powerful and sudden force pushing or tilting the animal is usually a predator, and the zebra fish doesn't want to stick around to find out. Furthermore, the direction of the behavioral response was always correct, with the animal always swimming away from the source of the current. However, the mutant zebra fish don't react. Without a working utricle, they cannot sense the perturbation, and they don't try to escape. In other words, Bagnall and colleagues discovered that the fish utricle is necessary for a very basic and critical behavior—the escape reflex.

The utricle and saccule are the oldest things within the inner ear, and each has a critical function. Yet, in clinical medicine they are largely ignored. Almost all vestibular tests examine the semicircular canals, and not the otolith organs. We aren't really sure how someone would react if their saccule stopped working, or the best way to retrain the utricle after damage. Theoretically, many of the symptoms that patients discuss could be traced back to the utricle and saccule. People frequently feel "floaty" and "ungrounded" or note that they veer to one side while walking, or that the world is sometimes tilted. All of those things would be expected when the gravity and tilt sensors go wonky. But we don't have good ways to test the otolith organs. The most commonly used otolith test, called a VEMP (short for Vestibular Evoked Myogenic Potential) test, doesn't directly test the utricle and saccule's primary functions. Instead, it uses loud sounds to activate them, which causes some measurable reflexive muscle activity. Moreover, many healthy individuals over the age of fifty don't have a VEMP response, making interpretation murky. I am curious as to whether we are on the precipice of a new chapter in vestibular medicine, where a greater understanding of the role of the otolith organs in health and disease helps us better diagnose and treat our patients.

THE UNSEEN WORLD

Now that we have met the hair cell, it's time to see it. Below are images courtesy of Nicolas Grillet, a professor at Stanford. I'm proud to include them in this book, as they are the best images of the hidden world of the hair that I have ever seen. They were taken with a scanning electron microscope, which is necessary to see the infinitesimal world inside the inner ear in vivid detail.

We see light, and you can see smaller and smaller things with magnification, but there is a physical limit to that magnification. Visible light is a wave, ranging from roughly 400 to 700 nanometers. Anything smaller will hide in the shadows between rays of light. To dive further, you need an electron microscope. Electrons are much smaller than light waves, and provide visualization at the level of proteins, the molecular machines inside cells. To see even further, into the quantum realm, you would need Ant-Man.

In other words, to see something, the probe has to be smaller than the object being studied. A crude analogy—let's say you are in a dark room, and you are asked to map out an unknown object. Now, the object happens to be a space-themed LEGO monorail set, which you happily possess because your mom didn't throw it out when she downsized from house to beachside condo so that your neighbor could bulldoze your childhood home to build a tennis court. But you don't know that. You have to map it out, in the dark, either by touching it with a log, or the tip of a pen. I know, this analogy keeps getting weirder, but bear with me. Also assume that you are in terrific shape (go you!), and picking up the log is no issue. But, because the surface of the log is so big, there is a limit to the information you will get by using it to probe the monorail. You might mistakenly conclude that the unknown toy is a G.I. Joe aircraft carrier. On the other hand, the smaller size of the pen will give you much more information about shape. You now correctly identify the monorail, thereby defusing the bomb that I forgot to mention and saving the world. Still with me? Light waves are like the log in this analogy. To see the infinitesimal, a scanning electron microscope uses a beam of electrons, which are one hundred million times smaller than light waves. This allows an amazing view of an unseen world.

Now that you've marveled at these crystal-clear pictures, let's discuss them. These images are all taken from mice during scientific experiments. Each panel has a white bar at the bottom, which provides a sense of scale.

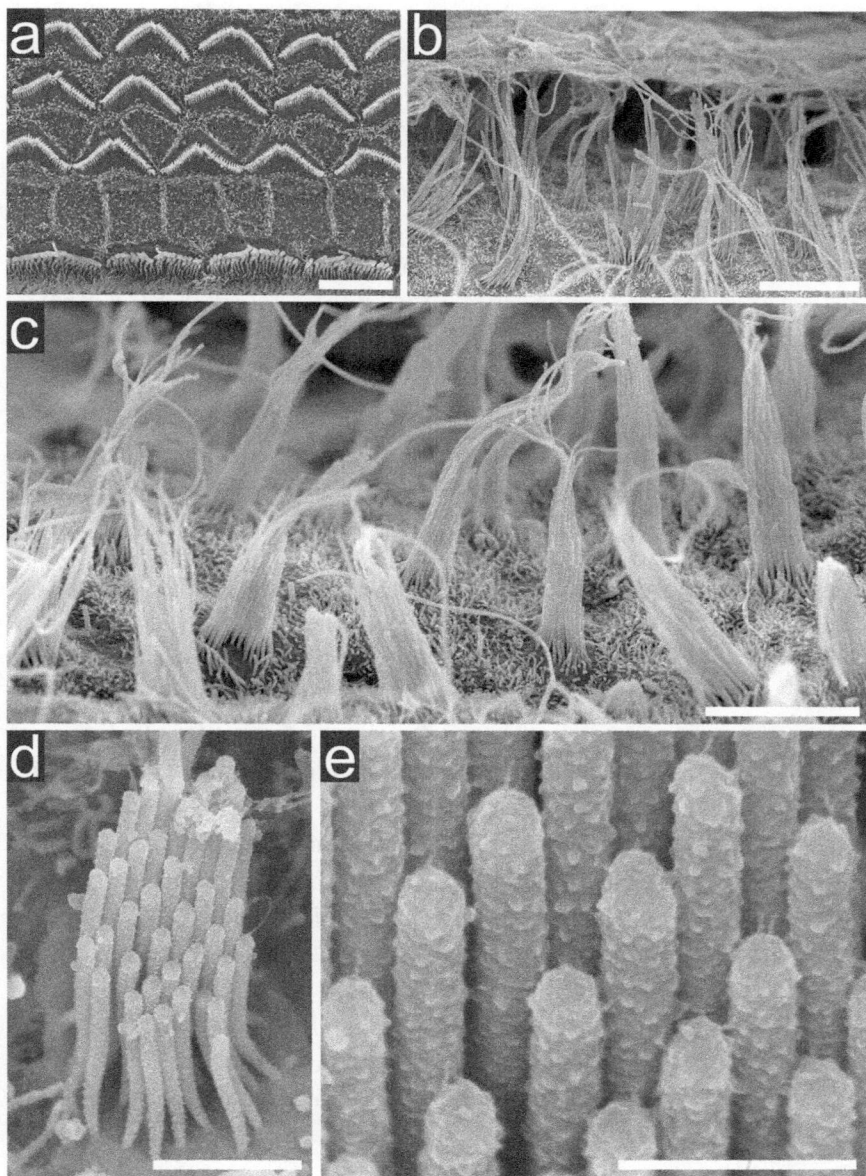

FIGURE 3.3 Scanning electron microscope images hair cells, part 1. (Courtesy of Nicolas Grillet)

For panels A through C, the white bar is five microns in length. My math fiends will recall that a micron is 10^{-6} meters, and that therefore there are a thousand microns in a millimeter. Panels D and E are more zoomed-in. The white bar in panel D is one micron, and the white bar in panel E is four hundred nanometers. Nanometers (10^{-9} meters) are the next level down in size after microns, so there are one thousand nanometers in a micron.

Panel A shows the inner and outer hair cells in the cochlea. The inner hair cells are the single row at the bottom, and the outer hair cells are the three rows at the top. As you can see, the hairs (stereocilia) are "V" shaped for outer hair cells, and in a straighter line for the inner hair cells.

Panel B shows hair cells in the otolith organs. Bundles of stereocilia protrude out of the top of each cell. You will recall that the longest stereocilia is called the kinocilium, and in this image you get a sense of how long the kinocilium is. At the top of the image, the stereocilia are embedded into a sponge-like layer. That is the otolithic membrane that we learned about, and it's the bed for the ear crystals. Looking at this image, it's easy to imagine how sliding movements of the otolithic membrane would pull on the stereocilia, opening their ion channels.

Panel C shows hair cells in a semicircular canal. You can see how each cell has a clustered bundle of stereocilia, neatly arranged by height, as though they knew that it was photo time! Furthermore, they all seem to be facing the same direction, with the shortest stereocilia closest to the "camera," and the tallest ones further away. This polarity is important because it lets the hair cells respond to specific directions of movement.

In Panel D, Dr. Grillet takes us deeper into the quantum realm. Instead of seeing a whole forest of hair cells, we now just focus on a single tree. Shown is a utricular hair cell, with all its stereocilia. The stereocilia are filled with a protein called actin, which provides structural support.

Panel E is the most magnified. Adjacent shafts of stereocilia stand tall, like downtown skyscrapers. Each one appears to be about one hundred nanometers across, about the size of the coronavirus that caused COVID-19. With its astounding magnification, this image shows a neat feature of stereocilia that is rarely seen. If you look closely, you can see small connections between adjacent stereocilia. Dr. Grillet explains that these nanofilaments connect the tip of a shorter stereocilia to its taller neighbor. These are the tip link proteins, and as the stereocilia sway with head movements, they yank open ion gates, triggering the cellular electrical cascade that quickly transforms to the nerve signal of movement, zapping to the brain.

Panel A shows several ear crystals. With this picture, you can really appreciate the crystalline structure of otoconia. They have a cylindrical shape, with sharp, tapered ends. The white scale bar is four microns in this image.

In Panel B, we are inside a semicircular canal, looking down on the crista, the saddle-shaped structure that serves as the scaffold for all the hair cells

FIGURE 3.4 Scanning electron microscope images, part 2. (Courtesy of Nicolas Grillet)

in the ampulla. Stereocilia bundles are seen protruding from the surface of the crista. The white scale bar is ten microns in this image.

VESTIBULAR MAGIC

As a child (and who am I kidding—as an adult too) I loved looking at visual illusions. Little tricks designed to expose the quirks and inner workings of the visual system. These trompe l'oeils are fun but also provide insights into a system that is absolutely remarkable in its capacity for scanning and processing the visual world but is certainly not a pixel by pixel transcription. This discrepancy is perhaps best illustrated by the visual blind spot illusion. We perceive a seamless world. However, each eye has a literal blind spot, where nothing is seen. Anatomically, it's where the optic nerve connects to the retina, which is the photosensitive area at the back of the eyeball. So, why don't we see a blank area? First, we have two eyes, so each eye covers the other's blind spot. Second, the blind spot is located in our peripheral vision, so you can only see it when you are looking at something else. Third, our brains lie. They fill the blind spot with information from the surrounding area, just like when I Photoshop out unwanted elements from photos. So to see the blind spot, using figure 3.5, cover your left eye, line up your right eye with the tree, focus on the tree, and move your head slowly closer to the page. At some point, the hungry lion just disappears—poof!

Are there any vestibular illusions? In a visual illusion, our sense of sight is tricked into giving incorrect information. With a vestibular illusion, our sense of balance would be tricked into giving incorrect information—about our movement or orientation. It turns out that the vestibular system is really good at providing accurate information during most of our daily activities. But, like any biological system, it can be fooled.

As we saw earlier, with head rotations, the cupula is displaced from its normal position, which is the cardinal event that allows for sensation of head

FIGURE 3.5 A visual illusion showing our blind spot.

turns. So, anything that moves the cupula will cause that sensation. With sustained rotations—like spinning around in place—at first the cupula moves because of the acceleration. But if you keep spinning, you will reach a steady speed, and you'll stop accelerating. It's like a car—you get pushed back into your seat during the acceleration from 0 to 60 mph, but while traveling at 60 mph, there's no longer any acceleration. At constant spin, after about five or six seconds, the rubbery cupula has snapped back to its normal position. If you suddenly stop spinning, then the endolymph in the semicircular canal will move the cupula with its inertial force. Just like in a car, if you slam on the brakes, the car stops and you lurch forward against the seatbelt with your inertial force. (This is an example of Newton's first law of motion). So, even though you have stopped spinning because of the cupula mechanics the brain gets fooled into thinking that you are twirling in the opposite direction. You are not moving, but you think you are. Voila!—vestibular illusion.

There's another interesting variant of the illusion where you can trick your brain into thinking that you are moving in an unexpected direction. It also involves spinning around in place. (I should formally state that I advised you to try this in a safe environment in case you fall). But, this time, instead of keeping your head upright, tilt your head so that it's touching your shoulder. With your head sideways, spin around in place. Stop suddenly and bring your head back up to its normal position. I've personally tried this one ad nauseum (literally . . .). You feel like you are moving, but up and down (the direction depends on whether you spun clockwise or counterclockwise). You'll either feel like you are falling through the floor, or in the initial stages of a backflip. How did the vestibular system get fooled this time? Again, it comes back to the cupula. When spinning around with your head on its side, you are activating the cupulas of either the superior or posterior canals. The head experiences the same relative movement as it would if you were spinning around in a human hamster wheel. When the movement stops, you still get the inertial cupula bounce for the vertical canals. Because the head is now upright, that's interpreted as a strong, sustained upward or downward motion. The vestibular system was duped! Not only were you not moving upward or downward, you weren't moving at all.

But it's not all fun and games. Our vestibular system evolved to provide accurate information for bipedal apes, living a terrestrial existence. The operating range of the vestibular system, where it functions best—with a gain of one—is from ~0.05 to ~5 hertz, which also happens to be the range of speeds for everyday head movements.[11] It also works best in tandem with the visual system, so that errors in perceived orientation or tilt can be rectified. In normal life, it functions so flawlessly that many people have no idea that they even have a vestibular system. But aircraft pilots need to be very aware of their vestibular systems (it needs to be on their radar . . . sorry, couldn't help it, forgive me). In poor visibility conditions, vestibular misinformation can cause the pilot to think that the orientation of the aircraft relative to the ground is wrong. Planes can move in ways that are confusing for inner ears—like slow banked turns—and which are often below the sensory threshold of the vestibular system. A number of vestibular illusions can occur—like the leans, or the ominously named graveyard spin (which is the aerial version of the spinning in place illusion explained above) and even more ominously named graveyard spiral. If the pilot trusts their ears instead of their instruments, the results can be disastrous.

The leans is thought to be the most common vestibular illusion experienced by pilots. It usually starts with a gently banked slow turn, which is not sensed by the pilot, because the turn is so subtle that it's under the threshold of perception. The pilot notices that the plane isn't straight—because of instruments—so he or she straightens the plane. That roll is quicker, so it generates enough force to be sensed by the vestibular system. From the aircraft's standpoint, it was tilted, and now it's straight. From the pilot's "incorrect" perspective, it was straight, and now it's tilted. So, the pilot may erroneously try to roll the plane to falsely correct course. Vestibular illusions are more common than most people realize, especially for small aircraft, and they are implicated in many crashes during cloudy, stormy, or foggy conditions. Given the thick fog at the time, it's been theorized that a vestibular illusion—a spatial disorientation—may have contributed to the helicopter crash that killed Kobe Bryant, his daughter, and seven others in January 2020. In the pilot's last communication, he indicated that he thought the aircraft was rising, to get above the cloud layer, when in fact the helicopter was already plummeting to the ground.

Another feared vestibular distortion is the somatogravic illusion. Recall from above Einstein's uncertainty principle. It states that the physical force produced by gravity and by inertia are indistinguishable. The somatogravic illusion involves the utricle, and ambiguity that is a direct result of its design. When you tilt your head backward, to look up at the ceiling, the crystal layer sitting on the utricle slides backward due to gravity, bending the hair cells below. If you accelerate rapidly in a straight line, like, say, a fighter pilot being catapulted off the flight deck of an aircraft carrier, the same thing occurs. The crystals slide back—due to inertia—and bend the hair cells backward. The hair cells can't tell whether the head is tilted back (gravity), or whether you are rapidly accelerating forward (inertia), because the physical effect on the hair cell is the same. This uncertainty can trick the brain. During acceleration, the pilot can be fooled into thinking that their head is tilted back, when it isn't. Since they can see the cockpit, perceptually, they know that their head can't be tilted back, so they assume the whole plane is tilted backward. In other words, they think the plane is flying nose up, rather than straight. It's a powerful illusion, and pilots have crashed because of it, as they try to correct by tilting the nose of the plane downward. In level flight, especially right after takeoff, this is catastrophic, because the plane and the ground quickly collide. Other senses can correct the false sensation, which is why errors and crashes mostly occur with poor visibility, like at night or in dense fog. Pilots are taught, time and time again, to ignore their faulty vestibular sensations, and rely on instruments.

In this chapter, we focused on the hair cell. Their core function—sensing movement in the microscopic environment around them—seems simple enough. However, our inner ears have taken advantage of hair cells to construct several remarkable sensors that underlie our ability to balance and hear. The function of each sensor explains the design of different parts of the inner ear: the spirals of the cochlea, the interlocking rings of the canals, and the crystals of the otolith organs. In future chapters, we'll cover what the brain does with all that information, and also what happens when different parts of this remarkable machine begin to malfunction. But for now, we can just marvel at the beauty of the inner ear, our elegant solution to sense the physical properties of the world around us, hundreds of millions of years in the making.

4

The Eyes Have It

Vestibulo-Ocular Reflex

THE BLINK OF AN EYE

We are protected by reflexes, an invisible shield of hardwired responses to lurking dangers. The cough reflex is the guardian of the windpipe, expelling aspirated food and liquid. The acoustic reflex protects ears from splitting sounds by stiffening the eardrum. Sneezes are the gatekeepers to sinus passages, flushing out unwarranted particles in a gust of air. Eyelids are a force field, snapping shut to cover the delicate surface of the eye.

To be effective, reflexes must be quick. Incredibly quick. A second can be split into 1,000 milliseconds, which is a good scale for discussing reflexes. Conscious thought takes a minimum of ~150 milliseconds. You can test this yourself on websites like humanbenchmark.com. A green screen is shown, and you click when it turns red. It took me about 300 milliseconds to respond, which may explain why my professional sports career never materialized. But, for most reflexes, by the time you realize something is wrong, it's too late. The reflexes must occur without thinking, with simple neural circuits that connect a stimulus (water in the windpipe) to an action (cough). A protective blink, triggered by touching the eye, takes about one hundred milliseconds. But for the vestibular system, the blink of an eye isn't fast enough. The vestibulo-ocular reflex, which keeps vision in focus during movement, can take as little as five milliseconds.[1] It's the quickest known reflex in the human body. Compared to the blink reflex, this isn't a race of the tortoise versus the hare; it's the tortoise versus a Lamborghini.

How does the vestibulo-ocular reflex achieve lightning speed? Researchers Marko Huterer and Kathleen Cullen break it down. As the head begins to move, hair cell stereocilia are bent, sending an electric current down the hair cell, which signals the vestibular nerve to fire. That takes 0.7 milliseconds. There's a three-neuron relay to cross the distance between the ear and the eye. Each neuron passes the electric signal to the next, like runners handing off a baton. The quickest measured time is about 0.7 milliseconds per neuron. Finally, the last neuron in the chain precipitates a muscle twitch to move the eye, which takes 2.4 milliseconds. The vestibulo-ocular reflex is critical for survival—like, say, keeping prey in sight during a hunt—and its astounding performance highlights its primal importance.

Lorente de Nó, a Spanish researcher, mapped out the neural pathway of the vestibulo-ocular reflex. He had trained with Santiago Ramón y Cajal, an early pioneer of brain anatomy. Ramón y Cajal studied brain sections using a histologic technique developed by Camillo Golgi, which allowed clear visualization of nerve tissue under the microscope. Previously, it was very difficult to spot neurons and their complex web of connections. Golgi experimented with various ways of staining cells to make them easier to see with a microscope, and found that silver nitrate stained neurons beautifully, turning them black. Using Golgi's technique, Ramón y Cajal would then meticulously draw different types of neurons, like trees in silhouette, with branches and roots radiating outward from the central cell body. For their combined work, Ramón y Cajal and Golgi shared the 1906 Nobel Prize. Despite this shared honor, they were bitter academic rivals. Golgi believed that the brain consisted of a vast continuous network called the reticulum. Ironically, Ramón y Cajal used Golgi's technique to disprove his own theory: the new stain made it clear that nerve tissue wasn't one massive, interconnected cell, but instead billions of interwoven individual nerve cells.

Lorente de Nó spent time with Robert Bárány, who we previously met as the only person ever to win a Nobel Prize for vestibular physiology. Lorente chilled with Bárány in Uppsala, Sweden, where he combined his histologic prowess with Bárány's physiology know-how to accomplish a number of pivotal experiments. In addition to showing that the vestibulo-ocular reflex arc only involved three neurons, he also made several observations on how the brainstem processes vestibular information. He discovered that the brainstem didn't just take in the vestibular signal and send it out unchanged.

Instead, he found that the brainstem amplified the signal. This amplification makes the vestibular signal last about three times longer than it would otherwise. Today, we call that feature "velocity storage." Its main purpose is that at low speed, the natural functioning of the vestibular system isn't strong enough to move eyes sufficiently to keep them locked on a target. Therefore, by amplifying the signal, velocity storage makes the vestibular system able to control eye movements accurately over a wider range of rotational speeds.

Back to the vestibulo-ocular reflex. The eye is a biological camera. Rays of light enter the eye through the pupil (aperture), get focused by the lens (lens . . . duh!), and are projected onto the retina (photosensor/film). But the retina is different from standard film, which is equally sensitive to light in all areas. In the retina, about half the photoreceptors are concentrated in a small cluster called the fovea, about 1.5 mm wide. The other half of photoreceptors are dispersed throughout the rest of the retina. Therefore, visual acuity drops off steeply as you move further out from the fovea. In order to effectively use the fovea, our eyes are constantly darting around, scanning the world. What we think we see is actually a mental reconstruction of a picture from its parts, like puzzle pieces coming together. To keep vision sharp, eye muscles are constantly at work, aligning the tiny fovea with visual targets. If you picture the visual world as a 360-degree sphere encircling the body, the fovea only covers a swath of one degree. Because of the fovea, control of eye position must be incredibly precise.

There are several eye muscle control systems that keep the fovea centered on visual targets of interest. They are complementary, with each control module regulating a different type of eye movement. Together, they enable clear vision in all circumstances, both at rest and in motion. The four primary types of eye movements are called smooth pursuit, saccade, optokinetic, and vestibulo-ocular reflex (for those in the know, yep, I've decided to ignore vergence. I am already using too much technical language! Ok, fine, since you must know, vergence allows cross-eyes when you are trying to look at something really close to your nose. In all other eye movements, the eyes are kept parallel, but to see something up close, each eye must face a different direction). With smooth pursuit, slow moving objects are continuously traced across the visual field. Saccades (from the French word for twitch) are quick, darting eye movements that redirect gaze to an object of interest. Optokinetic movements move eyes to keep up with a moving visual field—like when everything you can see is moving. Since we normally move around the world, and not the other way around, optokinetic activation can result in a compelling illusion of movement. The classic example of this is when a train passenger watches another train rolling into the station, occupying most of their field of view, and they mistakenly believe that they are moving. And finally, the vestibulo-ocular reflex adjusts eye position to compensate for head movements. Optokinetic and vestibular control can be thought of as complementary, as they both try to react to head movement. Optokinetics work best during slow head movements because they rely on visual processing which does take some time. Vestibulo-ocular dynamics are suited for quick head movements, which generate higher acceleration forces against hair cells. Together, they help ensure that our vision can remain steady throughout the natural range of head movements encountered during everyday activities, like skydiving onto a snowmobile, or kicking a soccer ball.

To visualize the difference between smooth pursuit and the vestibulo-ocular reflex, try this: hold out your index finger at arm's length, and move it side to side. (Interesting aside—unlike saccades, smooth pursuit is not under voluntary control. You need a target to track to smoothly move your eyes. Try as you might, without something to focus on, you can only dart your eyes around with saccades). By moving your arm slowly at first, and then speeding up, you can find the speed at which your finger blurs. That is the upper speed limit for smooth pursuit, which is about fifty degrees/second. Now, try the opposite: instead of moving your finger, shake your head side to side while staring at a stationary finger. This task uses the vestibular reflexes, and you'll find that your finger remains sharp with much quicker movements. At high enough speeds, like two hundred degrees per second, only the vestibular system can stabilize vision.

SEEING EYE TO EYE

Vision without the vestibulo-ocular reflex is like a homemade movie from the 1980s, as your dad runs over to capture your triumphant excursion down the neighborhood slide. The scene bounces around erratically, making it hard to see anything. The retina, just like the film, is subject to motion blur. In order to cancel out bobbing, eyes must counteract each head movement with an equal and opposite eye movement. Head jerks up, eyes jerk down. Head tilts right, eyes tilt left. Head snaps left; eyes snap right. If the ratio of head movement to eye movement is one, then net movement is zero, and the high definition fovea stays right on target. The vestibulo-ocular reflex is a neuromechanical solution to eliminate image shake.

There's a surprising symmetry between the inner ear and the eye socket. Eye movements are controlled by six muscles. Each muscle has a partner in the same plane, to allow for movement in either direction (muscles can only get shorter, not longer. Therefore, muscles are always arranged in pairs. As one contracts, the other is reset to a position where it can again contract). The three planes in the eye correspond to the three planes of the semicircular canals. The horizontal canal shares the same orientation as the medial rectus and lateral rectus muscles. Same for the right superior canal and the right superior and inferior rectus muscles, and the right posterior canal and the right superior and inferior oblique muscles. This symmetry is necessary for the speed of the vestibulo-ocular reflex. Our brains are capable of complex calculations (quick, multiply 7×12!), but it takes time. Each additional neuron in a circuit adds to the tally. If the canals and muscles were not aligned, mental arithmetic would be required to accurately translate canal signals into correct eye movements. Conversely, the coplanar design ensures that geometric information is hardwired, enabling a lean three-neuron circuit to accurately move the eyes.

There are times when it doesn't make sense to have a vestibulo-ocular reflex (VOR). If you are trying to track a moving object, like a flying baseball, you want your eyes to move with your head, not in opposition to it. The VOR is designed for situations where you and your head are moving, but the visual target is not. When the target is moving, we use other strategies to keep it in focus, including moving our heads, and the smooth pursuit eye movements that we learned about earlier.

To be able to appropriately help when needed and inactivate when not required, the VOR needs an on/off switch. We call that switch "VOR cancellation." It's located in the cerebellum (which we will cover more in the next chapter), and it can be damaged. Loss of VOR cancellation causes a peculiar problem: difficulty with tracking moving objects. It's easy to test for that in the clinic, by having your patient hold out one arm, and fixate on their thumb. The examiner then rotates the exam chair. With normal VOR cancellation, eyes ignore the vestibular signal arising from the rotating chair and stay glued to the thumb. Without VOR cancellation, the brain cannot ignore the rotation, and the VOR causes nystagmus, even though the eye twitching results in a blurry thumb. So, VOR cancellation is an essential ability that allows us to use the VOR when advantageous and shut it off when not.

MAKING A HIT

In 1988, an article was published in the *Archives of Neurology* that would change vestibular medicine forever. It was called *A Clinical Sign of Canal Paresis,* and there were two authors: G. Michael Halmagyi and Ian Curthoys.[2] The article described what came to be known as "The Head Impulse Test." The test is deceptively simple. It only takes a minute to do, and no special equipment is required. In teaching trainees about the Head Impulse Test, I borrow the slogan from the board game Othello. The Head Impulse Test takes "A minute to learn, a lifetime to master." Sometimes we abbreviate the Head Impulse Test as the HIT. So technically, yes, I do HIT my patients. Don't tell anyone.

The Head Impulse Test solves a very basic problem. It allows you to quickly and accurately assess the VOR. Remember that the first step of the VOR is one of the semicircular canals sensing a head turn. With damage— like from trauma, infection, inflammation, or a tumor—the semicircular canal won't be able to initiate the VOR. The HIT lets you see that. Prior to the HIT, you needed expensive and cumbersome laboratory testing to assess the vestibular system. With the HIT, anyone can assess the VOR, anywhere, for free.

Here's how the Head Impulse Test works. An impulse is a quick head turn. The examiner and the patient face each other, with heads at the same height. The patient needs to fix their eyes on a target. Typically, the easiest thing to stare at is the examiner's nose. Next, the examiner takes hold of the patient's head and quickly turns it to one side. This part needs to be explained carefully, and I usually mimic the head turn in slow motion first to demonstrate what will happen. Based on movies, we all believe that it's very easy to just twist someone's head right off! Also, many people have neck problems or are sensitive to quick movements, so of course one must explain everything and ask permission before doing this test. The examiner follows the patient's eyes. With an intact VOR, the eyes will stay glued on the target/nose. That's because the VOR is working and counteracting the head movement with an equal and opposite eye movement. With a faulty VOR, the eyes will move the head and come off the target/nose. That's because the VOR didn't work and the head turn wasn't sensed, so the eyes just

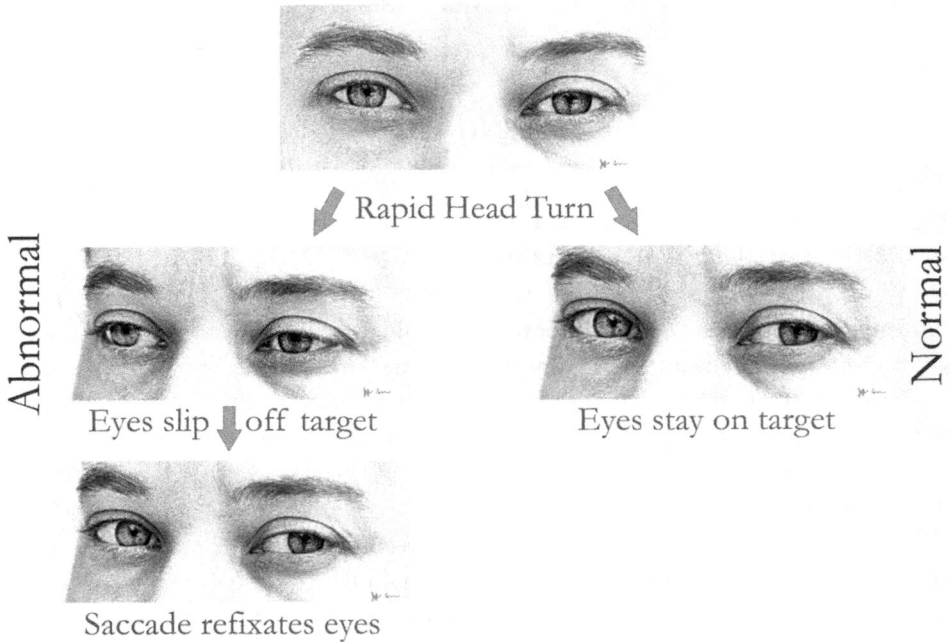

Rapid Head Turn

Abnormal

Normal

Eyes slip off target

Eyes stay on target

Saccade refixates eyes

FIGURE 4.1 The head impulse test.

naturally moved together with the head. A few hundred milliseconds later, the scene is visually processed, and the patient realizes that they are staring at the cheek, and not the nose. The brain then sends a signal for a quick eye movement to get the focus back on the target/nose. That is called the "catch-up saccade." So, if the examiner sees a catch-up saccade, then the VOR isn't working.

The Head Impulse Test can be done for each of the semicircular canals. So, there are really six HITs that you can do for each patient, assessing all the canals. There might seem to be a problem with that. Each of the six canals has a paired canal in the same plane. For example, when you shake your head side to side—"no"—you are in the plane of *both* horizontal canals. So, how does the HIT give you specific information about a single canal, when anatomically you are always stimulating two canals?

The answer is best explained by Professor Halmagyi himself. I interviewed him for this book, and I'm going to put his whole quotation down here. Not because I'm a lazy author, but because he's a brilliant man, and it's worthwhile following his train of thought, as he—and his longtime colleague Ian Curthoys—discover the single best vestibular test ever devised. Me being lazy has nothing to do with it:

The Head Impulse Test is based entirely on our knowledge of the physiology. And the key physiological experiment was done by Jay Goldberg and Cesár Fernández. They subjected monkeys, they recorded off the superior vestibular nerve, and they gave them constant accelerations . . . and of course, it's elementary. They knew what they were going to find. They found a resting rate in that neuron, say 80 spikes a second. So you could tell immediately that you could drive that neuron to 800 spikes a second. But you could only drive it down to zero. Neurons can't fire at a negative rate. So there has to be a asymmetry. Any neuron that has a resting rate is more excitable than dis-facilitatable. The common term is inhibition, but it's not inhibition. Inhibition is an active process with inhibitory neuron. This is not inhibition, this is decreasing the resting rate. And therefore, Ian and I concluded from that, that someone who has only one lateral canal, there has to be an inhibitory saturation. Right? And therefore, if you drive that neuron fast enough in the off direction, the VOR will saturate, it has to—no doubt about it. And therefore, if the person's awake, they have to make a compensatory eye movement, which can only be saccadic. And that is the catch-up saccade of the head impulse test. And the reason I knew about catch-up saccades was because I'd also trained in neuro-ophthalmology. And their cover test to see if someone's got a strabismus, depends on looking at the catch-up saccades. So combining cellular neurophysiology of the vestibular system with clinical neuro-ophthalmology gives you the answer. And you know the answer before you touch the patient. It has to be there. And the problem was, people have tried, but the mistake they always made is they turned the head too slowly. They did not reach inhibitory saturation. It was as simple as that. You do it fast enough, you see it. So, when you actually look at possibly the most lasting contribution we've made, head impulse testing depends on that. Fundamental cellular physiology and clinical observation. And you put the two together, bingo, you have something new.

Professor Halmagyi is a fascinating person. He's known around the world as a master clinician and innovator. During our interview, I saw some other sides of him as well. He's a natural storyteller, recounting Bárány's remarkable life that we explored in chapter 1. He's a world traveler and has doled out numerous recommendations for various countries. When traveling to Uppsala, where the Bárány Society meeting is held periodically, one must pay homage to Carolus Linnaeus's garden, where he devised our modern system of taxonomy. Halmagyi is philosophical in his answers. When asked how he came up with the Head Impulse Test, when no one else did, he quoted Goethe: "you can only see what you are prepared to see."

Halmagyi was no "One-HIT wonder." In his partnership with scientist Ian Curthoys, and engineers, they devised and validated the video Head Impulse Test. It's the same test, but the head turn is quantified with a movement sensor, and the eye movements are captured using a high-speed camera. The data is then processed with software. There is an advantage to the video HIT, compared to the original version. Our eyes are not able to see catch-up saccades that occur during the head movement. We can only see catch-up saccades that happen after. Right after an injury, the catch-up saccades occur pretty late—200 to 400 milliseconds after the start of the head turn. Because those late saccades are easy to see, they are called "overt saccades." But, with time, the brain compensates. It begins to predict head movement and prepares the corrective saccade ahead of time. So, the catch-up saccade gets earlier and earlier, happening between ~70 and 200 milliseconds. The quicker saccades in that range will occur during the head impulse. Those hidden saccades are called "covert saccades." Your eyes can't see them, but a high-speed camera can. There are now several commercially available video HIT systems out there, enabling easy diagnosis of vestibular damage. They also provide an approximation of vestibular recovery after injury but showing the latency (the time delay) of the catch-up saccade.

Halmagyi had other contributions as well—too many to list here. But I must mention one more: The VEMP test. Like all sciency things, it's an acronym. Vestibular Evoked Myogenic Potentials. I know. It's a mouthful. So, if you are thinking about skipping the rest of this chapter, Instagram is only a click away. But here's the counterargument: Instagram has basically zero VEMP content. I looked. There's nothing there. Cute puppies? Check. Cute kittens? Check. Cute puppies meeting cute kittens for the first time? Check, check, check. But VEMPs? Nada. So, you are stuck with me. Let's continue.

To understand the VEMP, which is a test of the otolith organs (utricle and saccule), we need to remember some biology and some physics. In humans, since we have a cochlea, the otolith organs are used for vestibular purposes—sensing gravity, tilts, and linear movements. But in other animals, the saccule and the utricle sense sounds. Human utricles and saccules—despite being vestibular organs—have retained some of this ability and will activate with very loud sounds. Furthermore, recall that part of the function of the utricle and saccule is to help our bodies balance. In order to do so, they are connected to muscle-controlling nerves throughout the body. So, with activation of the utricle and saccule, there are changes in muscle activity, which can be a stiffening or a loosening of the muscle in question. Sometimes these changes are visible—like a muscle twitch—but sometimes they can only be recorded with electrodes. Time to put it all together. A VEMP is an induced change in muscle activity, caused by activation of the utricle and/or saccule, and the triggering stimulus can be a loud sound. Hence, "Vestibular Evoked Myogenic Potential."

Building on earlier work by VEMP pioneer Reginald G. Bickford, G. Michael Halmagyi, James Colebatch, and Nevell Skuse described the cervical VEMP, or cVEMP.[3] Halmagyi told me that he enjoyed doing research with Colebatch partially because he was a "wonderful electrophysiologist," but also because both their fathers used to do research together, on heart disease. The cervical VEMP was measured from the sternocleidomastoid muscle, the "v" shaped cord in the front part of the neck, connecting a point below the ears to the sternum. In order to prove that the cervical VEMP response was vestibular (i.e., saccule or utricle) and not auditory (i.e., cochlear) in nature, the research team took a direct approach. In the '70s and '80s, it was common to sever the vestibular nerve for intractable vertigo. So, they included a group of patients who had had their vestibular nerves cut but still had normal hearing. In those patients, the VEMP response was abolished, proving it to be a vestibular and not an auditory reflex. Today, VEMPs are still used to assess the health of the utricle and saccule. And, unexpectedly, they were also found to be a great test for another disease we'll learn about in chapter 11: superior semicircular canal dehiscence.

It's time to say goodbye to this chapter. We learned all about the vestibulo-ocular reflex, which is the most basic function of the vestibular system. The VOR uses vestibular signals to keep vision steady during movement. And while the VOR is foundational, the vestibular system is so much more than just the VOR. Let's journey onward, into the mysterious realm of the brain.

5

The Brains Behind the Operation

NUCLEAR POWER

The vestibular system generates an incredible amount of information by monitoring head orientation and movement. That data is then sent to the brain along the vestibular nerve. Humans have 22,000 individual neurons in the vestibular nerve at birth, declining to about 16,000 by ninety years of age.[1] Countless tiny nerve fibers exit the back of the inner ear, through miniscule channels in the bone, and coalesce to form the nerve, like hairs woven together to form a braid. The nerve then crosses a centimeter-long tunnel through the base of the skull—the internal auditory canal. After emerging from the canal, the nerve crosses an open space called the "cerebellopontine angle" to join the brain. It's estimated that each second, each vestibular nerve sends over one million nerve signals to the brain! In this chapter, we'll see how the brain refines vestibular information to help with eyesight, balance, autonomic functions, alertness, and spatial thinking.

The vestibular nerve connects to a primitive part of the brain called the brainstem. The brainstem is ancient and integral. It's the brain's life support system, regulating breathing, heart rate and blood pressure, alertness, and sleep. In addition, all nerve fibers between the rest of the brain and the spinal cord/body pass through the brainstem. Finally, the brainstem controls the cranial nerves: twelve specialized fibers that connect to different parts of the head. This book is focused on only one of those nerves: the eighth. Others are responsible for smell (I), eyesight (II), moving the eyes (III, IV, and VI), sensation on the face (V), moving the face (VII), throat sensation (IX), control of the voice box (X), neck turning (XI), and tongue movement (XII).

Humans can survive damage to many parts of the brain, albeit with profound neurologic consequences. However, serious damage to the brainstem inevitably results in death. In fact, "brain death" is really defined as an irreversible and permanent loss of brainstem function. That's why the examination for brain death focuses on brain stem reflexes—like breathing, the gag reflex, or the caloric response (a vestibular reflex).

The vestibular nerve connects to four vestibular nuclei in the brainstem. A "nucleus" in the brain is a hub of interconnected neurons, capable of performing calculations, like a computer processor. A number of neural circuits radiate outward from the nuclei, sending out vestibular data to different parts of the brain. These hardwired networks allow the vestibular system to carry out its cardinal functions. Many patients with vestibular disease complain not only of dizziness, but also of blurred vision, instability, fatigue, confusion, and lightheadedness with positional changes. Sadly, these symptoms are frequently discounted by medical specialists, but they seem to follow logically from the neuroanatomic layout of the vestibular system.

Vestibular nerve fibers fan out from the nuclei to integrate with: (I) eye movement control centers, to steady vision in motion (oculomotor nuclei), (II) spinal cord nerve bundles that maintain balance by directing postural muscles (vestibulospinal tract), (III) multisensory integration center (thalamus), (IV) wakefulness and alertness area (reticular formation), and (V) memory and spatial cognitive center (hippocampus). In addition, the (VI) vestibular nuclei on each side of the head are connected to each other, which has important implications for our ability to compensate after vestibular damage on one side. Finally, the (VII) cerebellum (literally "little brain") has extensive connections to the vestibular nuclei, which help control and fine-tune vestibular responses. Numbers I, II, and V have each earned their own chapter (chapters 4, 6, and 7 respectively), so they won't be covered here. But fret not, there are still plenty of interesting things to discuss!

LOSING YOUR NERVES

The brainstem serves to organize and process incoming vestibular information. In the ear, vestibular information is simple: it's just the information directly sensed from each canal or otolith organ. But in the brainstem, that information gets sorted into streams of higher-level, processed information. While it might seem like a minor point, there are actually major implications of that anatomic distinction. Implications that emergency room doctors and neurologists use every day to try to figure out if a patient who suddenly developed vertigo is at risk of dying or not. The basic problem is that two diseases that are really different can clinically appear quite similar. Those two diseases are vestibular neuritis and stroke.

For reasons that are not totally understood, many of the cranial nerves will suddenly stop working. When the facial nerve (the seventh cranial nerve, controlling the muscles for facial expression) stops working, half of the face becomes paralyzed. We call that condition Bell's palsy, after Sir Charles Bell, a Scottish anatomist. The condition is devastating, causing a grotesque visage, with a lopsided smile, sagging skin, and unblinking eyes (for those affected—forgive my choice of words—I opted for overly poetic language to paint a grim portrait for dramatic effect). Those affected frequently experience social isolation and depression, having lost the ability to emotionally communicate through facial expression. Thankfully, for most, the condition is temporary. When the hearing nerve (part of the eighth cranial nerve) stops working, we call that condition "idiopathic sudden sensorineural hearing loss." From medicalese to English, this translates to "somebody lost their hearing, and we really have no idea why." The other half of the eighth nerve, the vestibular nerve, can suddenly stop working as well. That is called vestibular neuritis (neuritis translating to nerve inflammation).

Recall from chapter 4 that the vestibular nerves aren't quiet when at rest. Instead, they each have a baseline firing rate, sending a steady stream of pings to the brain. Picture a game of tug-of-war (and if you prefer a high-stakes version, picture the tug-of-war from *Squid Game*). At the onset, both sides are pulling equally. Even though there is a lot of force, there is no net movement, and everything is still. If one team stumbles, suddenly the pull of the other team is uncovered, and the rope gets jerked to one side. With vestibular neuritis, when one nerve gets inflamed and suddenly stops working, an asymmetry also follows. The healthy nerve is still sending the usual

~100 spikes per second to the brain. The unhealthy nerve is sending zero. From the brain's perspective, that's the same exact signal to the healthy nerve as would occur if the whole body was being spun around. That causes nystagmus (the rhythmic eye twitching) and vertigo (the powerful sensation of being inside a washing machine).

Stroke, defined as a death of neural tissue, can occur anywhere in the brain. Neurons die quickly without the oxygen and sugar that blood constantly provides. Stroke is the fifth most common cause of death in the United States. There are two main sets of blood pipes going to the brain. In the front of your neck, hiding under the jugular vein, are the carotid arteries. In the back of the neck, traveling through the spine, are the vertebral arteries. The vertebral arteries provide the blood supply to the brainstem and the cerebellum. So, if a stroke affects the vertebral artery, or one of its branches, then it can affect the vestibular system. Just like with vestibular neuritis, a stroke can cause nystagmus and vertigo.

Because brainstem vestibular data and ear vestibular data are stored in different formats, different nystagmus patterns result from injury. Recall that due to Ewald's Laws, nystagmus patterns aren't random. Instead, they reflect the specific activation of individual semicircular canals, or combinations thereof. For example, if Angelina Jolie suddenly loses all vestibular function from her right ear, then her alluring eyes will twitch to the left, mostly horizontally, but with a small twist as well. That's what you get when you add up unopposed activation of the three canals on the left. The horizontal canal says you are spinning like a top. The superior and posterior canals are more complex, each having a vertical vector and a twist (roll) vector. The vertical vectors just cancel each other out, so the eyes don't move up and down. The twisting is synergistic for both canals, so it does contribute to the final nystagmus output. So, with vestibular neuritis, we expect a horizontal pattern nystagmus with a rotary component.

With a stroke, it's no longer the raw data stream from the semicircular canals that is damaged, but processed data. For example, data from both posterior canals is pooled together in the upper layer of a region of the brainstem called the medulla. Lesions in that area knock out all posterior canal information, leaving the paired superior canals unopposed. Since the superior canals normally sense downward head movements (like in a forward somersault), the brain responds, with a nystagmus directed downward. Now, it's hard to think of a disease that would take out both posterior canals and leave the rest of the canals intact. Therefore, vestibular physicians know that when they see a purely vertical nystagmus, the problem is almost certainly in the brain. (Author's note: for the smarty pants out there. If you just destroyed a single posterior canal, yes, you would get a nystagmus with a downbeat component. But, because the canal is forty-five degrees off axis, you would also get a twisting component to the nystagmus. So, it wouldn't be purely vertical).

Therefore, nystagmus patterns are an incredible clue as to the cause of sudden onset vertigo and nystagmus. They help tell the difference between vestibular neuritis, a benign condition usually treated with steroids and physical therapy, and a stroke, which of course can be life-threatening. In fact, it's been shown that a careful examination of the eyes in the emergency room is more accurate than an MRI.[2]

DOING THE MATH

There's a basic design problem in the vestibular system. The inputs don't match the required outputs. The inputs, constrained by physics, encode information about acceleration. And, as I've no doubt said too often at this point, a large chunk of that data stream is used to keep your eyes steady as you move. And if you just faithfully transmitted the acceleration data to the eyes, they would countermove appropriately. At least at first. But the acceleration signal fades quickly, and eyes have a natural, elastic tendency to snap back to their starting position. For eyes to stay steady on an object of interest, they need to turn in proportion to the head movement and then stay put. The eye muscles can't just contract once, during the head acceleration. They have to keep contracting afterward, to counteract the viscoelastic forces of the eye socket pulling the eyes back to center. Raw vestibular input isn't good enough.

The brainstem has a solution to this problem, called the neural integrator. Now, I know what you are thinking: integrator . . . that rings a bell . . . isn't that from calculus? Well, you are right. No matter how well you did in high school math, as a fundamental fact, your brain can do calculus. Let's go back to school for a moment. The fundamental attributes of any moving object are position, velocity, and acceleration. It could be a baseball that you've tossed toward your mother-in-law, or a water balloon full of urine. Doesn't really matter. The object can be described by position, velocity, and acceleration. And those three are related to each other mathematically. Velocity is change in *position* over time. Acceleration is change in *velocity* over time. And calculus can be used to transition from position to velocity to acceleration. If you *differentiate* position, you get velocity, and if you *differentiate* velocity, you get acceleration. In the opposite direction, if you *integrate* acceleration you get velocity, and if you *integrate* velocity you get position. So yes, in case you are wondering, I was on my high school math team, but no, being a star mathlete didn't make me as popular as you would think.

Let's lay it out. The brainstem must convert the acceleration signal into a position signal. It needs to tell the eyes where they should be. That's where the neural integrator comes in. The acceleration data gets integrated (in the calculus sense of the word) until it becomes position data. By doing some math, the brainstem is able to ensure that the vestibulo-ocular reflex works perfectly every time. And since the integrator is located in a neural network, it can be disrupted. When it doesn't function properly, it's called a "leaky" integrator. One type of nystagmus, called "gaze-evoked nystagmus," is thought to result from a dysfunctional integrator. Without the integrator, when the eyes are looking off to one side, they can't maintain that position. Instead, they start drifting back to the center. However, the signal to look off to the side continues, resulting in a back-and-forth oscillation of the eye that is only present when looking to the side, and hence the name "gaze-evoked nystagmus."

THE LITTLE BRAIN

The cerebellum (literally "little brain") lies in the back, bottom part of the skull. It functions as a neural computer, fine-tuning complex and coordinated muscle movements. Whether training to bobsled, pole-vault, figure skate, or high dive, it's really the cerebellum that "learns." Without a cerebellum, smooth movement isn't possible. Neurologists have a simple test for the cerebellum: trying to touch the examiner's moving finger. Like many little things in life, it's a remarkable feat that normally occurs so flawlessly we don't even notice it. It requires an accurate computation of the finger's position in space, and then a complex and perfectly timed sequence of motor commands to move the arm properly. Trapezius, 32 percent activation to lift the arm; Bicep, 17 percent activation to bend the elbow; pronator teres, 54 percent activation to turn the wrist; extensor digitorum, 74 percent activation, to outstretch the finger. You can still move without a cerebellum, but movements are clumsy and uncoordinated. With a faulty cerebellum, movements degrade from ideal and efficient, into tremulous, erratic, and inaccurate zigzags.

To borrow from military terminology, the cerebellum calculates a "firing solution" for each movement, the same way that the operator of an artillery battery needs to calculate a firing solution to accurately hit a distant target with a projectile. And yes, astute reader, there's a pun in there because nerves fire as well as guns. Let's take an extreme example. At the onset of World War One in the summer of 1914, the German army steamrolled through Belgium, and advanced to within thirty miles of Paris. They were halted during the "Miracle on the Marne," beginning the era of gruesome trench warfare that characterized the next four years of conflict. The Germans, unable to advance, fabricated the "Paris Gun," a behemoth that bears the distinction of being the longest-range artillery ever created. Fired from behind German lines, shells flew up to a height of twenty-five miles, at speeds of 3,500 mph, to hit targets seventy-five miles away. The firing solution of our cerebelli is no less amazing, enabling wondrous feats of balance and coordination (like when I successfully toss garbage into an open can from five feet away).

The cerebellum isn't all about muscle coordination (called motor control). As we'll see, the cerebellum has a critical role in the vestibular system. Just like with body movements, the cerebellum fine-tunes, adapts, and modifies vestibular output to ensure accuracy. In order to study adaptation in the vestibular system, scientists have come up with some interesting experiments.

In chapter 4, we had a lot of fun exploring the vestibulo-ocular reflex (VOR). In order to keep eyes steady, for each head turn, there is an equal and opposite eye turn. We saw that the ratio of eye movement to head movement is called the gain, and normal gain is one. Starting in the 1950s, scientists started investigating if the VOR could learn. Under normal circumstances, it makes sense that for every one degree of head movement, there should be one degree of eye countermovement. That's because for each degree of head movement, the visual world (or your target) moves one degree as well. But what if you change that relationship? By using specialized optics, researchers have been able to probe for VOR adaptability. Experiments have been done with glasses and prisms, placed in front of the eyes, which magnify the visual world, minimize the visual world, or even reverse it, like a mirror image.

When the world changes, the vestibulo-ocular reflex (VOR) is able to adapt.[3] If you place 2× magnifying glasses on an animal, then every time the head moves, any object of interest will move twice as much. Picture how quickly the world moves when you look through binoculars. Because of that, in about a week of continuous usage, the VOR adapts. The gain goes up, so that one degree of head movement is accompanied by two degrees of eye response, so that the target is held steady. The opposite occurs with a minimizing lens. With that, visual targets will move slower than normal for each head movement, so the VOR gain drops in response. Steve Lisberger, an expert on cerebellar learning, has studied this extensively. He has shown that in monkeys, if you surgically remove two vestibular parts of the cerebellum, called the flocculus and paraflocculus, then the animals can't learn to change their VOR anymore. They are no longer able to adapt to the topsy-turvy worlds of magnification or minimization.

VOR experiments have been done on humans as well.[4] In a series of experiments in Innsbruck, Austria, research team Theodor Erismann and Ivo Kohler investigated the effects of world-reversing goggles. They would make research subjects, and themselves, wear the goggles for weeks on end. Some goggles, with a mirror, turned the world upside down, others reversed right and left. With an upside-down view of the world, they found that subjects stumbled and struggled for the first few days. "For instance, the participant held a cup upside down when it was about to be filled; or they stepped over a ceiling lamp or street sign." However, around the sixth day, the world perceptually became upright again. After that, subjects would resume normal activities, including walking, playing sports, and even riding motorbikes across town. When the goggles were removed, the world would again flip, and a few more days were needed for everything to go back to normal. When others have followed up on these funhouse experiments, they have found that VOR gain responds appropriately to each situation, even reversing when necessary (for prisms that horizontally flip the world).

The cerebellum isn't just helpful for learning in kooky laboratory experiments. If we lose vestibular function—because of an infection, a tumor, trauma, or any other cause—the cerebellum is critical in helping us relearn how to balance. Mutant mice without a cerebellum don't recover after a vestibular injury.[5] A major way the brain learns is through error signals, and trying to minimize them, so animals kept in the dark and animals with restricted movement don't compensate either.[6] After a sudden loss of vestibular function on one side, several problems emerge. The first is that the healthy nerve keeps up its baseline firing rate while the damaged nerve is quiet, and that asymmetry creates a powerful sensation of spinning (vertigo), and nystagmus. Over a few days, the cerebellum fixes that issue, first by driving down the firing rate of the healthy nerve, and then by assisting in restoring a baseline firing rate for the damaged nerve. That is called "static" compensation, because it helps with constitutive vestibular activity (activity that is happening all the time, even when you aren't moving). The cerebellum also helps improve vestibulo-ocular reflex gain and the timing of corrective eye movements, although due to physical limitations it never recovers to normal. Since that refers to processing occurring during head movement, it's called "dynamic" compensation.

Part of my job duties include partnering with neurosurgery to remove benign brain tumors called vestibular schwannomas. These tumors grow from the vestibular nerve, so the nerve is resected during microsurgical excision. The tumors are located partly in the skull, and partly in the brain vault, hence the two-surgeon team. When you test the vestibular system prior to surgery, most people have lost function, but some have not. Paradoxically, those who still have some vestibular function left feel worse after surgery, because they suddenly lose all residual function. After surgery, we'll see nystagmus, dizziness, nausea, and imbalance, because of the vestibular loss (of course, they do have some of those symptoms because they just finished a six-hour brain surgery as well). Over the next few days, we get to see the cerebellum at work, fixing the vestibular system and eliminating the nystagmus. Physical therapy helps the process, and multiple studies have shown that it improves recovery and balance after vestibular schwannoma surgery.

THE BIG BRAIN

Let's recap. Vestibular information is generated in the inner ear, and then travels down the vestibular nerve, into the brainstem. There, in the vestibular nuclei, signals are organized and processed. In addition, the cerebellum is connected to the vestibular brainstem, and helps with learning, adaptation, and compensation after injury. Nice. We are definitely going to ace the test now. But, for extra credit, let's keep going. The next stop for many vestibular fibers, after the brainstem, is the thalamus.

The thalamus is shaped like an egg, and it's located very close to the center of the brain. There are two, one on each side of the head. The thalamus collects data streams from sensory systems, including vision, touch, pain, temperature, auditory, taste, proprioception, and of course, vestibular. It then directs those streams to the proper cortical area for processing and perception. For our senses, the thalamus is the central train station, and most sensory signals must pass through it to get where they are going.

From the thalamus, vestibular signals converge on several regions of the brain's cortex. The cortex is the outer region of the brain. It's the reason why our brains have so many fissures and grooves, which function to make the cortex bigger by multiplying its surface area. The cortex is highly organized, with six layers of cells. It's where thinking, planning, and perception occur. Most sensory systems have a primary cortex—a central hub where neural noise is translated into a distinct sensation. In the visual cortex, located in the occipital lobe in the back of the brain, pixels from light- and color-sensing cells in the retina are understood. A dark background, punctured by mere photons of light, too many to count. A central swath of haziness, glowing, irregular—like a great scar arcing across the darkness. Neurons babble in excitement. Recognition, it's the Milky Way. The memory and emotion files are checked. It's beautiful. It's the earthbound view of our cosmic neighborhood, a spiraling eddy of hundreds of billions of stars. It's special. It makes us feel small and alone, in a vast universe.

Each sensory system has its own primary cortex. In that area, all the neurons react to stimuli from that sense. Hearing has the auditory cortex, located in the temporal lobe. Touch has the somatosensory cortex, smell the olfactory cortex, and taste the gustatory cortex. The vestibular system, conversely, has no primary cortex. Neural mapping and neuroimaging studies have shown that there are cortical regions where many neurons are sensitive to vestibular information. These areas are spread widely across the parietal, temporal, frontal, and insular regions of the brain. The exact function of those areas is only beginning to be understood. However, it seems that vestibular information is usually combined with information from other senses, like vision and somatosensation, to form estimates of heading, gravity, and position.

For example, one of these cortical areas—called the medial superior temporal area (MSTd)—appears to be important for estimating self-motion. In order to navigate any environment, you have to have an internal estimate of your heading. The MSTd is wired to both the visual and vestibular systems. That means that neurons in the MSTd will fire if vestibular input indicates movement in a certain direction, *or* if the visual system indicates movement in a certain direction. There are many neurons in the MSTd, and each one has a specific job, only firing for specific trajectory of movement. For example, neuron 172,324,975,234 (aka Bob the Neuron) might only fire if you run bearing five degrees to the right. Dora Angelaki, a neuroscientist, has studied the MSTd for years, trying to decipher the secrets of multisensory integration. Her team records directly from MSTd neurons in monkeys as they traverse a virtual environment. Movement is conveyed through optic flow—when the entire visual field, everything you can see, moves—and also through vestibular stimulation, as the entire experiment is conducted on a mobile platform. Angelaki has found that by combining visual and vestibular information—multisensory integration—the MSTd is a navigational computer, constantly updating estimates of self-motion.[7]

Lesions in those cortical areas (e.g., from a stroke) cause errors in those estimates of heading, gravity, and position, resulting in several bizarre symptoms. With the aptly named "room tilt illusion," the internal estimate of gravity is incorrect, resulting in a perceptually flipped or slanted world. In "pusher" syndrome, the world isn't tilted, instead it's the patient. They try to correct this illusion by leaning over, in an effort to align their bodies with the erroneous perception of uprightness. Finally, with thalamic astasia, a person cannot walk, despite healthy muscles and intact sensation. While the exact mechanism is not precisely understood, it occurs after damage to the part of the thalamus that relays vestibular information, and therefore it may reflect an acute inability to balance due to invalid estimates of body position and orientation.[8]

In an effort to fully catalog all the connections between the vestibular nuclei and the rest of the brain, a Chinese research team took a comprehensive approach. Wanting to trace connections between neurons, they turned to nature for answers. The world's deadliest virus—by far—is rabies. Without a timely vaccine, rabies is 100 percent lethal. If you are infected with Ebola or smallpox, there is a chance that you will survive. Not so with rabies. Inoculation occurs with a bite from an infected animal. Once inside, the virus locates the closest nerve cell, and hijacks cellular transportation mechanisms to hitch a ride along the axon to the brain, hopping from neuron to neuron. The research team exploited this, by using a mutant virus imbued with a luminescent tracer to infect mouse vestibular nuclei.[9] The mice were then sacrificed (a commonly used research euphemism), and their brains were thinly sliced, analyzed, and used to construct a 3D model of every brain hub connected to the vestibular nuclei by just a single neuron.

The researchers found extensive connections between the vestibular system and other parts of the brain. For example, as expected, they found that the vestibular nuclei were wired to the oculomotor nuclei, which is how vestibular signals are sent to the eye movement centers, necessary for the vestibulo-ocular reflex. Two interesting findings were highlighted. First, a link was found to several brain areas involved in the circadian rhythm—the daily sleep/wake cycle controlled by a biological master clock in the hypothalamus. Data is sparse, but there may be a connection between vestibular dysfunction and impaired sleep. Additional links were found to two important brain nuclei: the locus coeruleus and the dorsal raphe nucleus. These two areas mediate stress and anxiety through two critical neurotransmitters: norepinephrine and serotonin (respectively). In fact, they are each the brain's major source of each neurotransmitter. You may recognize their names: most antidepression and antianxiety drugs work by modulating levels of those chemicals. Furthermore, many antidepression and antianxiety drugs are used in the treatment of dizziness. Benzodiazepines (like Valium) are given as a vestibular suppressant to reduce vertigo, SSRI (selective serotonin reuptake inhibitors) are used in treating Persistent Postural Perceptual Dizziness (PPPD, a cause of chronic dizziness), and a variety of antidepressants (e.g., tricyclics like nortriptyline and mixed reuptake inhibitors like venlafaxine) are used to treat vestibular migraines.

The connection between anxiety and vestibular disorders is very common. In my own practice, about half of those with dizziness and/or vertigo suffer from anxiety. Sometimes, patients don't want to discuss their anxiety, worrying that all their symptoms will be blamed on it, preventing them from getting the help they need. I personally think the strategy that is most helpful is to acknowledge anxiety when present, but to recognize that it's almost never the primary cause of the dizziness. It should be treated, just like any other contributing factor. The hardwired connection between vestibular nuclei and the norepinephrine and serotonin centers in the brain may explain why we see anxiety and dizziness together so often. Furthermore, disorientation is a powerful sign that the body is in danger, leading to an activation of the threat response system. In the short term—for example, the altered sensorium you didn't expect because your brother swore that these brownies were safe to eat—that's a good thing, a protective mechanism to keep you from harm while temporarily disabled. But, in the long term this response can be counterproductive, with continuous red alerts from the threat response system precipitating and perpetuating anxiety.

There's a constant frustration in vestibular medicine. Patients can't seem to describe their symptoms. They know something is off and will try to relate their symptoms with umbrella terms like "dizziness." Doctors, taught to be precise, try to probe deeper. "Please explain how you are feeling, without using the word dizzy." Patients will sputter, racking their brains to put words on their unease. Both are disappointed: the patient feeling that they are squandering precious minutes with the doctor, who doesn't appear to understand, and the doctor, not quite sure how to proceed. It's only by studying neuroanatomy that we can understand why. It's hard to describe vestibular sensations because there is no single vestibular cortex. It's a subconsciousness and deeply integrated circuitry, and therefore quite difficult to specifically describe the feeling of vestibular dysfunction.

We don't even have a word for our ability to feel vestibular sensations. We "see" with our eyes. So, I can tell my ophthalmologist that I can't see anything, or what I see is blurry, or that I have double vision. But what is the vestibular verb? We incorrectly state that we "balance" with our ears. But as we have seen, balance is a multisensory feat. It's more precise to say that we sense acceleration with our ears. But there isn't a single word for that. I can smell, hear, touch, or taste. I cannot "vestibulate." "Accelerate" has a different meaning. Language has failed the vestibular system.

6

Don't Fall for It

Vestibulospinal Reflexes

CATLIKE REFLEXES

I had a hamster growing up. Perhaps as the first indication that I was headed for medical school, I named him "Doogie Hamster," after the precocious, fictional, teenage physician, Doogie Howser. I had wanted a puppy, but this request was rejected by my parental units. Doogie eventually escaped from his cage, and presumably hopped on a train going north, and lived out his days in a hamster colony in New York City, enjoying avant-garde art shows. Years later, I adopted another pet. Puppy was still off the table, as I was working thirty-hour shifts during my intern year. I wanted something larger than a hamster, cuter than an iguana, and more interactive than a goldfish. With those guiding principles, I adopted a cat. After much reflection, he was named Snoop Catt, in honor of the great musician Snoop Dogg.

It was therefore with some element of horror that I discovered that cats were a favorite experimental animal for vestibular researchers. In 1922, Rudolf Magnus, a German physiologist, published "Wie sich die fallende Katze in der Luft umdreht," or "How the Falling Cat Turns in the Air."[1] In that experiment, cats had their vestibular system surgically removed—a bilateral labyrinthectomy. The cats were then suspended in the air, upside-down, and dropped to see if they could right themselves. Healthy cats are remarkable acrobats, quickly twisting their bodies to land on their feet. Without a vestibular system, the cats "flop to the ground like a sack." A cat's ability to tell that it is falling and compensate depends on having an intact vestibular system. The vestibular system isn't just important for stabilizing vision. It's also important for stabilizing the whole animal.

A working vestibular system helps prevent falls. In a study of older adults, Doctor Yuri Agrawal and colleagues measured vestibular function, vision, grip strength (a commonly used metric of general strength), and somatosensory function.[2] They then built a mathematical model correlating all those variables with falls over the previous five years, controlling for age and sex. Remarkably, when all the variables were considered together, vestibular dysfunction was the only significant predictor of falls, raising the odds of having a fall five-fold.

This finding has huge implications for public health efforts. Consider that 10 percent of older adults fall at least twice annually, that 30 percent of community dwelling adults fall at least once annually, and that 10 percent of falls result in serious and life-threatening problems, like hip fractures. Yearly, this results in ~3 million ER visits, and healthcare costs of about fifty billion dollars.[3] It's clear to me that any effort to promote healthy aging must include an assessment of the vestibular system, and a treatment plan with directed physical therapy when an issue is found.

In 1936, under the title of "Honorary Aurist, Bootle General Hospital," Alex Tumarkin put forth an argument to try to explain a rare manifestation of Ménière's disease.[4] As we'll see in chapter 9, Ménière's disease is a degenerative condition of the inner ear. Attacks of severe vertigo are the most salient feature, but hearing loss, tinnitus, and ear pressure are all common as well. However, a small percentage of patients suffer from "drop attacks." Drop attacks are scary. They occur without any warning, and result in a loss of muscle tone and balance, resulting in a quick fall to the floor. Many patients have described it to me as though they were shoved to the floor violently by some unseen force. Importantly, there is no loss of consciousness, which helps distinguish a drop attack from other causes of sudden falls.

Tumarkin suggested that the drop attacks of Ménière's disease were due to surges in otolith output, affecting postural tone, which is the baseline muscle activity that keeps us upright. With this apoplexy of the otolith organs, "The patient will double up and collapse backwards like an empty suit of clothes." Tumarkin called the drop attacks "The Otolithic

Catastrophe," however today they are commonly known as the "Otolithic Crisis of Tumarkin."

FALLING SICK

Proper balance is a team effort, with contributions from your eyes, sense of touch (feeling the ground beneath your feet), and vestibular system. Your brain then uses that information to make adjustments to postural muscles, keeping you upright. But, even when other senses are removed, by, say, closing your eyes and standing on a pillow, most people can still balance pretty well. That's because the inner ear vestibular system is sensing each sway, tilt, or turn, and can direct an appropriate postural countermeasure. How does the vestibular system accomplish that feat?

Vestibulospinal reflexes, as the name implies, refer to neural connections between the vestibular organs and the spinal cord, which serve to control postural muscles. Roughly speaking, we have extensor muscles, which generally increase the angle of any joint, straightening the back, legs, and arms. Extensor muscles are activated while standing. Flexor muscles do the opposite, curling us up in the fetal position when activated. The vestibulospinal reflexes help control the balance of tone between extensor and flexor muscles.

Vestibulospinal reflexes aren't just important for maintaining balance. They are also critical for adjusting the body during a fall, to minimize damage. There are several strategies for falling safely, including protecting the head by bending the neck forward, trying to land on cushier areas like the buttocks, and using arms and legs in a semibent position to help break the fall. In 1975, Richard Greenwood and Anthony Hopkins concluded a study looking at muscle responses during experimental falls in a laboratory. Research subjects were harnessed, pulled up by a pulley, and suddenly dropped without warning from a variety of heights.[5] Greenwood and Hopkins found that an initial burst of muscle activity in the calf was present in all test subjects, except those without vestibular function. Not only are those who suffer from vestibular disorders more likely to fall, they are also more likely to have a bad fall because they don't brace properly.

In addition to helping with postural muscles, the vestibular system has another mechanism for helping with balance. As we move, it's important to keep our heads as upright as possible. Head upright is the usual orientation for all information coming into the brain's command center, and as anyone who's tried to watch TV upside down will attest. Therefore, the vestibular system has a mechanism—the vestibulocollic reflex—to help keep our head steady during movement by reflexively moving neck muscles. This reflex is important for humans. It's also really important for birds, who need to keep their heads steady during the turbulence of winged flight. Even while denied vision, and no matter the axis of body rotation, birds are able to keep their heads remarkably stable. For those interested, there are several videos on YouTube demonstrating this remarkable feat.

We actually take advantage of the vestibulocollic reflex in the clinic, to test the vestibular system. If you put electrodes on the sternocleidomastoid muscle in the neck, and activate the vestibular system, you can measure electrical changes due to muscle activity in the sternocleidomastoid muscle (I know the muscle name is a mouthful, it describes the three attachment points of the muscle—the mastoid, the sternum, and the clavicle). This test, called the VEMP—Vestibular Evoked Myogenic Potentials—was first described by James Colebatch and colleagues in 1994.[6] Further research confirmed that the reflex originated in the saccule. Interestingly, the easiest way to elicit a VEMP response is with a loud sound, because when sounds are loud enough, they activate the saccule, in addition to the cochlea. Recall that some animals use the saccule to hear, not balance. VEMP is a really useful test to check on whether or not the saccule is working. And there's another version of the VEMP test, called the ocular VEMP, that tests the utricle. Combined with a video head impulse test (vHIT), which allows testing of each semicircular canal, VEMPs allow for assessment of all five parts of the vestibular system.

Conceptually, the vestibulocollic reflex (VCR), vestibulospinal reflex (VSR) and vestibulo-ocular reflex (VOR) all serve slightly different purposes. They all rely on the vestibular system to sense head movements. But nature is clever. Once it has a technology, it finds creative ways to put it to use. The VOR exists to keep vision steady, despite the fact that our heads and bodies may be moving. The vestibular system senses movements and counteracts by moving eyes to compensate. The VCR exists to keep heads steady, despite moving bodies. The vestibular system senses movements of the body and counteracts by directing neck muscles to move the head, canceling out the head movement. The VSR exists to keep our bodies steady. The vestibular system senses head and body movements, and directs body muscles in our arms, legs, and core to maintain balance and recover from falls.

The relationship between the vestibular system and coordinated muscle movements may be more complex than many realize. Researcher David Schoppik conducts experiments on zebra fish, which are small, striped minnows endemic to the waters of South Asia.[7] Normally, as zebra fish develop after birth, they begin to coordinate the movement of their bodies and their pectoral fins to efficiently swim upward. In mutant fish with deficient vestibular systems (who are otherwise entirely normal), they discovered that the zebra fish were unable to coordinate body and fin movement. They were able to swim efficiently with their bodies, and they were able to thrust effectively with their pectoral fins; they just could not perform both actions synergistically. Schoppik concluded that during a critical early period of growth, feedback from the vestibular system is necessary to guide the development of complex, coordinated movements. The zebra fish need to sense whether their bodies are upright or not to provide feedback during the motor learning process, especially for more complex movements. Without the vestibular information, the animals can't assess if they can maintain balance during movement, which hinders progress. Schoppik points out that this may be more important in aquatic creatures, as land animals might be able to substitute somatosensory information (like increased weight on one foot versus the other) while learning complex motor behaviors (like say hopping around on one foot).

The vestibulospinal reflex is clearly important for maintaining balance. In the Introduction, I presented the traditional line of thinking, where vestibular information is combined with visual and touch signals to enable balance. One of my mentors—Doctor Tim Hullar—has been challenging that standard view. Tim is an intellectual, a scientist, and a friend. He was also one of my inspirations for choosing my particular career path. I vividly recall doing a cochlear implant as a resident under his guidance. After the case, he turned to me and said that he thought I could make it as an ear surgeon. Gaining the trust of those we admire is really important, and it's something I try to keep in mind as the cycle of life has progressed, as I mentor my own trainees.

Doctor Hullar has been trying to prove that in addition to vision and touch, hearing is also important for balance. In one experiment, he found that research subjects were able to balance better in the presence of an external sound. Balance was measured with sway on a force plate—so it's a test of how still you can stand. With the sound on, sway was reduced by 41 percent. Our brains are constantly updating a mental representation of the world around us—a virtual version of our immediate environment. The results argue that just as visual input is used to update that model, auditory information is helpful as well. We mentally construct a soundscape, and it's another orienting clue that the brain can use to improve balance. Hullar is the first to admit that hearing is not as important as vision or touch sensation for maintaining balance. Still, when trying to prevent falls, every little bit counts. In a follow-up study, his team studied the next question.[8] If hearing is important for balance, then does restoring hearing improve balance? To answer that, the team had subjects balance on a foam platform, with a speaker generating white noise in front, like an acoustic anchor. Balance was then assessed with and without hearing aids, and they found significant improvements when the participants were wearing hearing aids. Corroborating evidence for the idea that hearing can improve balance comes from another researcher, Dr. Sharon Cushing, who has found in a series of experiments that cochlear implantation—another form of hearing restoration—improves balance in children.

In this chapter, we saw how the vestibular system contributes to balance, and how it may be the single most important sensory system for preventing falls. In the next chapter, we are going to probe deeper into higher and more mysterious levels of brain function—memory, navigation, and finally the ethereal realm of thought itself.

7

A Balanced Mind

Memory and Cognition

FAR-OUT THINKING

Our discussion in previous chapters was *grounded* in the science of how the vestibular system gathers information. And *we came close* to having a complete picture of the purpose of that information. But now it's time to *put that behind us* and *move forward* into *uncharted territory.* It's time to *raise big* ideas, while *staying away* from pure conjecture. But not to worry, I shall try to be a dutiful guide, so that *we don't get stuck.* We need a *360-degree bird's-eye view* of the vestibular system. Simply stated, it's time to *think outside the box.*

Ok, ok, I think y'all can see what I did there. All the italicized words are metaphors that use space to describe the substance and process of thought. But is that just a neat trick of linguistics, or is there something deeper here? In this chapter, we will explore vestibular connections with memory, emotion, and abstract thought. We'll see how the brain combines vestibular information, along with visual, auditory, and somatosensory input to form a mental image of our place within the space around us. And we'll go *one step further* (sorry, can't help myself now), and discuss a fascinating idea: that this mental representation of space is integral to our ability to think abstractly, and perhaps necessary to the very essence of consciousness itself.

Let's set up a *road map* for the chapter. In order to figure out the connection between vestibular disease and thinking, we will look at several lines of evidence. First, we will see what happens in animal experiments when the vestibular system has been removed. Second, we will consider data from humans who have diseases that have affected their vestibular system. And finally, we will discuss the neural circuitry that connects vestibular, memory, and reasoning centers in the brain, and how that relates to our ability to engage in abstract reasoning.

Animal experiments are invaluable to modern science, enabling scientists to test ideas in a realistic biological environment. They allow us to intervene in ways that would not be possible in humans, like intentionally destroying parts of the body. This type of experiment is important, because it allows us to infer not just the association of events, but also causal relationships. There is a natural tendency to conflate *association* with *causation*. How many times have you seen a report on the evening news suggesting that you should drink more coffee or eat less chocolate to prevent cancer? Almost inevitably, if you investigate further, you find that in a large database, drinking more coffee was *associated* with less cancer. But, the referenced study is almost never interventional, and the research team didn't divide participants into groups who received different amounts of coffee per day in a controlled environment for thirty years, while noting the differential incidence of new cancers across groups. If they had, you could argue that if the groups were otherwise similar, then less coffee *causes* cancer. But if the two are just *associated*, then it's irresponsible to conclude anything further. Perhaps those who drink more coffee are wealthier and have better healthcare, which helps prevent cancer? There are a number of logical relationships between two *associated* things that don't involve *causation*.

In a series of experiments, Professor Paul Smith sought to figure out the effects of removing the vestibular system by embracing the scientific cliché of studying rats in a maze.[1] The rats had their vestibular systems in the inner ear surgically removed and were then allowed to heal for five weeks. They were then placed in a radial arm maze, which is a type of maze that resembles spokes on a wheel. There is a central circular platform, and eight arms extend outward from the central platform. At the end of each arm—not visible from the center—is an area where food pellets can be hidden. So, in each trial, one of the arms hides a food pellet. Lighting was provided for orientation. Finally, the food pellet was placed in the same arm each time. Therefore, this is a test of spatial learning and spatial memory. It's reminiscent of when I look for my keys in the morning, where I sequentially check each likely site, trying not to duplicate efforts to increase my odds of getting to work on time. Could the rats use their environment to orient themselves in the maze, and then remember which pathway would reward them with that sweet, sweet food pellet?

The experiment was structured like a test. In order to pass the test, the rats had to do the maze with either one error (one wrong turn) or zero errors (going straight to the arm with the food pellet). And they had to do that several times in a row, to eliminate the effects of a lucky guess. The normal rats did well, and all of them passed the test in three weeks. However, at that three-week mark, only half of the rats without a vestibular system were able to pass the test. The natural criticism of this type of study would be that without a vestibular system, perhaps the rats were too unsteady on their feet to complete the task. But that is not what Dr. Smith saw. The vestibular-deficient rats were still able to run around and get to wherever they were going in a reasonable amount of time. In fact, they didn't take any longer than the normal rats to complete the task. Instead, their issue was that they could not seem to remember where the food pellet was (spatial memory), or how to get there (spatial navigation). Destruction of the vestibular system in the inner ear didn't just cause a problem with their ability to balance. It caused a problem with their ability to think.

In a follow up experiment, rats were evaluated on their ability to find their way home after foraging for food in a maze. To remove the effects of visual cues and landmarks, this test was performed in the dark. Compared to normal rats, those without a vestibular system performed poorly, and after setting out from their home and eating, they were unable to find their way back. Their navigational guesses were essentially random—just like my wife's. The authors surmised that normal rats were using their vestibular system as a compass, measuring their relative direction of movement. By combining that information with how far they had gone—as measured by steps—they would be able to construct a mental map of where they had gone relative to their starting point. However, without their internal compass measuring their heading, the vestibular-deficient rats simply had no idea where they had come from. Like in the prior experiment, they were still able to move around and guess where their homes were, so it wasn't a question of balance. Also, they had been given five months to recover from their surgery. Instead, by removing their vestibular system, the scientists had impaired the rat's ability to navigate in the dark.

Why does damage to the inner ear affect spatial thinking to such a marked degree? To get to the bottom of this, we'll have to learn about how the brain keeps track of space, and our position within that space. The story starts with a name that is known to every neurologist in the world: HM. (Author's note: that this is different from H & M, which is a clothing store. Pay attention!). Born in 1926, HM suffered from epilepsy, which caused horrible seizures, unable to be controlled with medications. With hopes of returning to a normal life, HM underwent brain surgery in 1953. During surgery, a large portion of the temporal lobe was removed from both sides of HM's brain. This included brain regions known as the amygdala, the entorhinal cortex, and the hippocampus. The surgery was quite helpful in controlling the seizures, but it left HM with an unexpected complication: he was unable to form new memories. His short-term memory was intact, as was his motor memory (the ability to learn new tasks that involve coordination of muscle movements, like juggling). It should not escape our attention that computers also separate short-term memory in RAM from long-term memory in the hard drive. HM could not form new memories for almost anything else. Given that he was born in 1953, and died in 2008, in theory he could watch the original *Star Wars* trilogy every week for his last twenty-five years and be surprised about Darth Vader's real identity every time. (Forgive me, the profound implications of HM's condition are depressing to contemplate, so I chose to go with a lighthearted example). Over the course of his lifetime, and beyond (his brain was donated to science, and is available for viewing online), HM contributed immeasurably to our understanding of how specific brain regions handle memory. And the main location of our brain's hard drive was found to be in the hippocampus.

CAMPUS LIFE

This discovery about the hippocampus inspired a generation of scientists, who sought to uncover its secrets. John O'Keefe was one of those young investigators, and in the early 1970s he performed a series of groundbreaking experiments. Rats were implanted with a recording electrode, placed into single cells inside the hippocampus. Despite this invasive setup, the rats were still awake and allowed to move around their environment. Unexpectedly, they found a cluster of cells that would fire each time the rats would go to a specific area. These cells were called "place cells." If the rat was in my apartment, for example, then one place cell would fire if the rat was in the Skittles drawer, a different place cell would fire if the rat was in the gummy bear cupboard, and a third place cell would fire if the rat was in the chocolate closet. In essence, place cells create a virtual representation of the world around us inside the hippocampus. Like Google Maps, the current location is marked. In a hugely influential book, *The Hippocampus as a Cognitive Map*, O'Keefe referred to this neural network as a "cognitive map."[2] So, the hippocampus is important not just for memory, but also our internal GPS. For his work, O'Keefe was awarded the Nobel Prize in 2012.

In addition to place cells, there are head direction cells. If place cells are the map, then head direction cells are the compass. So, a place cell fires when you are in a particular location, and a head direction cell fires when you are facing a particular direction. Researcher Jean Laurens, a neuroscientist at Baylor University, studies head direction cells.[3] Others had already shown that head direction cells were sensitive to direction in 2D. North, south, east, and west. But the world isn't two-dimensional. Laurens was curious if head direction cells also encoded information about upward and downward tilt to form a three-dimensional encoding of head direction. To figure that out, he put rats on a platform that could freely rotate in any direction within a sphere, and then he recorded activity from different head direction cells throughout the brain. He found that some head direction cells sensed horizontal rotation, others sensed up and down tilt, and some sensed both. He also found that vision wasn't needed for the cells to orient (by repeating the experiments with the lights off). He concluded that vestibular input provides the raw data necessary for head direction cells—the neural compass—to orient us in three dimensions.

Is our cognitive map of the space around us—located in the hippocampus—affected by the vestibular system? In a 2003 study on rats, researchers found that the place cells stopped functioning normally once the inner ears were surgically removed.[4] Instead of precisely firing when the animal was in a particular location, cells would fire sporadically in many locations, as though the cell was no longer able to figure out where it was. The cognitive map had been broken. And the data showed that the cognitive effects of vestibular disease aren't limited to rats. It also affects humans.

In a landmark 2005 study, Professor Thomas Brandt and colleagues in Munich examined the hippocampus and spatial memory in patients with vestibular damage on both sides of their head.[5] They found that compared to normal control subjects, those with vestibular damage had measurably smaller hippocampi. Keep in mind that when an area of the brain is damaged, or loses its input, it will shrink in size. Use it or lose it! Furthermore, they also tested spatial thinking by assessing performance in a virtual maze. In the maze, subjects had to use reference landmarks outside of the maze to orient themselves, like sailors navigating by the stars. Similar to rats, humans with vestibular damage did worse in the virtual maze. They frequently headed in the wrong direction and couldn't seem to orient themselves. This two-part experiment showed that damage to the part of the brain responsible for spatial thinking has real world implications.

If the hippocampus shrinks without vestibular input, does anything make it grow? Eleanor Maguire, a hippocampal expert, sought to answer this question.[6] She compared the size of the hippocampus between experienced London taxi drivers and control subjects. The study was published in the year 2000, before Uber, Lyft, and GPS in general. In those days, a London taxi driver had to memorize every street name within the vast labyrinth of Western Europe's most populous city. London, founded over 2,000 years ago, is not a neat grid; it's a vast and complex patchwork of stately avenues, narrow paths, and side roads crisscrossing at odd angles, bisected by the serpentine River Thames. To memorize London, a taxi driver must construct one of the largest cognitive maps imaginable. Maguire found that this cognitive demand resulted in a measurement change in the size of the hippocampus. Taxi drivers had larger hippocampi than controls. Interestingly, the increased volume of the hippocampus was correlated with taxi driving time: the longer someone had been driving London's streets, the bigger their hippocampus was.

There's another curious bit of evidence that the vestibular system helps your brain understand where it is in space. French researchers Christophe Lopez and Maya Elziere found that while 5 percent of the public had experienced an out of body sensation, 14 percent of those with a vestibular disorder had—far higher than you'd expect.[7] Going further, a Swiss team found that they were able to induce an out of body experience with electric brain stimulation.[8] Of course, the research team wasn't just poking around the brain with an electric probe for fun; they were mapping brain activity prior to an epilepsy surgery. They found that with low-intensity stimulation of the right angular gyrus (an area on the border between the temporal and parietal lobe), the patient would report vestibular sensations, like a feeling of falling or "sinking into the bed." With higher-intensity stimulation, she would report a distinct out of body experience, where she felt she was floating above herself. The authors concluded that that the eerie feeling was "a result of failure to integrate complex somatosensory and vestibular information." If one function of the vestibular system is to keep you grounded and understand where your body is in space, it does make sense—however bizarre—that those certain patterns of vestibular dysfunction could cause you to feel "ungrounded," i.e., that you are outside your own body.

The connection between vestibular disease and cognition has been extensively researched by Dr. Yuri Agrawal, a surgeon-scientist at Johns Hopkins. I had the fortune to learn from Dr. Agrawal during my fellowship training. She's fiercely intelligent and deeply motivated. In one study, her research team took advantage of a large set of publicly available data that is collected annually by the National Health Interview Survey.[9] This study included a whopping 20,950 participants. Of those, 8.4 percent of U.S. adults were estimated to have vertigo (sensation of spinning or other movement) due to vestibular dysfunction each year. That's almost one in ten people. Compared to the general population, that group had an ~8-fold higher risk of "serious difficulty concentrating, remembering, or making decisions."

There are multiple lines of evidence that the vestibular system wears with age, like eyesight or hearing. The good news is that most people retain enough function to allow them to go about their lives, without barriers to their mobility, work, hobbies, or social activities. It is clear, however, that for each passing year, we should be spending more time deliberately training our sense of balance. Useful activities include yoga, tai chi, tennis, pingpong, walking, jogging, skiing, and anything else that involves balance and movement.

Since brain function can decline with age, Dr. Agrawal and her team looked at aging, vestibular dysfunction, and cognitive abilities. They wanted to see how often age-related brain problems are due to vestibular disease. This study was nested within a larger study, called the Baltimore Longitudinal Study of Aging. Participants were given a battery of cognitive tests, including some that required spatial thinking, and a vestibular test that assessed the saccule. Those with vestibular dysfunction did worse on four of the tests, which I'll describe. (1) In the card rotation test, you are shown a picture of an abstract shape. Next, you are shown a series of pictures of the same shape, only that it's been rotated a random amount each time. The catch is that sometimes, the rotated shape is not the same as the original shape, instead it's the mirror image. You must determine which shapes are the same as the original shape, and which are mirror images of the original shape. So, the card rotation test assesses how well you can mentally manipulate an object. (2) The Purdue Pegboard test assesses how many pegs can be placed into little holes, lined up on a board, in thirty seconds. It looks like you are trying to build a miniature bookshelf. The number of pegs placed within the time limit is correlated with manual dexterity, and hand-eye coordination. (3) In the Benton Visual Retention Test, you are shown a card with two or three printed geometric designs (like a square bisected by a vertical line) for ten seconds. You then have ten seconds to draw the design from memory. Ten cards are shown in each test. Therefore, it's a test of short-term memory for shapes (and artistic ability!). (4) Finally, in the Trail Making Test (Type B), small circles are spread randomly across a piece of paper. Each circle contains either a letter or a number. Without lifting up your pen, you have to connect the circles in order, alternating between numbers (1, 2, 3) and letters (A, B, C). Time and error rate are used to calculate the score. Across a variety of mental tasks involving spatial thinking and memory, people with vestibular dysfunction did worse. Using complex statistical models, the research team calculated that vestibular dysfunction was responsible for ~10 percent of the poorer scores seen with aging.

In another test of thinking, subjects are asked to mentally rotate objects. For example, in one version of the test, you are shown a drawing of a hand and asked to decide if it's a right hand or a left hand. To increase complexity, the palm or the back of the hand can be displayed, and additionally the hand can be rotated on its side or upside down. Both reaction times (how long it takes to figure out which hand it is) and error rates (how often you guess the wrong hand) are measured. A team of Swiss researchers were curious as to how those without a vestibular system would perform on this test. On the one hand, the test is done with the subject sitting still. Therefore, there is no direct vestibular information used, because both the body and head aren't moving. On the other hand, the test involves mental space, and the ability to visualize and manipulate objects within that space. As we've seen, vestibular input is critical for spatial reasoning. The research team found that those without a vestibular system on either side (bilateral vestibular loss) performed much worse than those with a normal vestibular system. Reaction times were significantly longer, and error rates significantly higher.

Dr. Agrawal took her research further. If the vestibular system is necessary for cognition, and if function declines with age, then is vestibular dysfunction associated with dementia? Dementia, sadly, needs no introduction.

We are all familiar with this sad end to beautiful life, when familiar faces go unrecognized, cherished memories are erased, language degenerates, and people get lost in their own homes. Dr. Agrawal provided some background on cognition:

> I think cognition can be broken down into about five domains. Language, attention, memory, executive function, and visuospatial or spatial processing. And the idea behind those different cognitive domains is that they are thought to be subserved by different neuroanatomic circuitry, although there's some overlap between them, but they sort of lead to different specific discrete kinds of behavior. And so the domain of cognitive function that seems to be most related to vestibular function, or that seems to rely most on vestibular information, is visuospatial cognition. And that encompasses specific skills and activities like spatial navigation, or the ability to move in a goal-directed way from one location to another, and spatial memory, which is the ability to record elements of one's environment in one's memory. So it's different from memories of events that occurred or memories of people. It's more a memory of, you know, you're visiting your childhood bedroom and you could remember exactly in your mind what the layout is.[10]

She then continued to explain her research into dementia and the vestibular system. I made the editorial decision to include a longer quote here, partially because writing a book is hard and copying and pasting is easy, and partially because her explanation is fascinating, and it reveals her thought process, which is worth following:

> So one of the questions that we sought to answer in observing a relationship between vestibular function and cognition, as part of healthy aging, was bringing this to its logical conclusion, looking at individuals who have the end stage phenotype of cognitive impairment, specifically Alzheimer's disease. And we were wondering, does vestibular loss lead to the cognitive impairment that occurs in patients with Alzheimer's disease? In Alzheimer's, there are multiple types of cognitive skills that degrade. And it turns out that they can sometimes degrade differentially within individuals. So some individuals may have the classic impairment in memory or episodic memory, that specific domain of cognitive ability, but may have preserved attention. It may be a disproportionate decline in one

domain versus another. There is a population of patients with Alzheimer's disease who have disproportionate declines in their spatial skills. They tend to fall, wander, become disoriented, lose track of things, have difficulty driving, have a number of specific types of deficits that are more spatial in nature, but maybe have preserved ability to remember individuals in their lives or things that have happened.

So the first question we asked was, is there a relationship between those patients who have that phenotype, the more spatial phenotype of Alzheimer's disease versus the ones who have the more classic Alzheimer's disease? And we found a huge difference that the patients with Alzheimer's disease who have more of the spatial phenotype have; nearly all of them have evidence of vestibular impairment versus the patients who have more of the classic phenotype, it's about a quarter who have evidence of some vestibular impairment, and that's about the level of what you'd expect just with the general aging process. So it seems like vestibular loss has one of its manifestations in patients with Alzheimer's disease: an advanced end-stage phenotype of cognitive deficit of spatial impairment. And it manifests as wandering, falls, disorientation, et cetera. And amongst the symptoms that individuals with Alzheimer's disease experience, those are amongst the most consequential and impactful. A person is wandering, you know, is not, you know, considered to be safe in their home environment. That's one of the primary drivers for institutionalization of those participants in caregiver burden.

As there is a connection with vestibular function, that is something that we're exploring in terms of prevention. If we offer vestibular therapy to patients with Alzheimer's disease who have evidence of vestibular impairment, does it improve some of those symptoms, such as falls, disorientation, and does it improve more specific measures of their spatial cognitive abilities as well? Vestibular therapy is typically not offered to patients with Alzheimer's disease. I think there's some idea, I mean, it's not on many people's radar screen in general, vestibular therapy, you know, even amongst healthy older adults. But I think there's some sense that patients with Alzheimer's disease have impaired motor learning as part of their neurodegeneration. And a few studies have suggested that they are capable of that, of motor learning, just with variations or modifications of the therapy techniques. And so where there's more participatory, there's caregiver involvement. And so we have a clinical trial currently looking at vestibular therapy in patients with Alzheimer's disease to see if it mitigates some of the negative outcomes like falls in patients with Alzheimer's disease.[11]

Without realizing it, all new parents use vestibular stimulation to comfort babies. Babies are soothed when rocked back and forth. My dad reports that I would calm instantly during "Rockabye Baby," and squeal joyously when dropped a foot as the "bough" broke. My mom reports that my dad messed up twice during my early childhood, once abandoning me in a hot car, and once letting me drink an unknown quantity of cleaning solution stored under the sink. Why am I telling you this? Because, dear reader, you deserve to know everything. But back to the story. To-and-fro swaying also comforts the elderly, perhaps explaining the perennial popularity of rocking chairs. We start life and end life with an innate preference to have our vestibular systems tickled.

And it's not just us. Apes, our genetic cousins, love to spin around. The other four species of great ape—chimpanzees, bonobos, orangutans, and gorillas—have all been observed whirling around for fun. Researchers Marcus Perlman and Adriano Lameira took advantage of YouTube to study videos of apes spinning, and to quantify the speed and duration.[12] They found that, on average, apes rotated at 1.4 revolutions per minute and lasted 5.4 revolutions. If you try this experiment at home, you'll find that speed is pretty fast. The authors hypothesized that the apes were trying to twirl themselves into an altered state of consciousness, and they highlighted other research suggesting that apes will also "consume fermented foods with alcoholic content." In other words, apes know how to party! As an aside, apes engage in a whole variety of other very "human" behaviors, including deception, problem-solving, building alliances, using tools, engaging in oral sex, trading favors for sex, playing, laughing, tickling, and forming gangs to murder rivals. You know, the usual stuff. The research team then compared apes spinning to human pinnacles of the pirouette: Ukrainian Hopak dancers, Sufi whirling dervishes, ballerinas, and circus performers. They found that the human apes and the other apes revolve at roughly the same speed. While the research team may be correct that apes enjoy the sensory altering aspect of vertigo, similar to the mind-bending trip of drug usage, it also seems possible to me that the apes just enjoy the vestibular stimulation, like kids on a merry-go-round.

Since we are on the topic of Sufi whirling dervishes: dervishes are expected to spin around for up to an hour as part of a mystical ceremony to entomb the ego, embrace the universe, emit love, and experience Allah. Sometimes, as well, there is another purpose—to entertain tourists. We have to wonder: How in the world can they whirl like a top for that long without getting dizzy? Once again, science may have an answer (you might be noticing a theme here). A team of scientists from Turkey (where the sect originated in the thirteenth century) scanned the brains of dervishes and normal controls using MRI to look for differences. They found that dervish brains were different, with thinning observed on the outer layer of the brain (the cortex) in several areas responsible for motion perception, including the precuneus, the right dorsolateral prefrontal cortex, the right lingual gyrus, the left visual area 5, and the left fusiform gyrus.[13] Therefore, the thinning can be understood as a self-imposed atrophy designed to downplay the natural function of that brain region. Visual areas 1 through 5 are part of the neural circuitry that processes visual information, and area 5 is responsible for visual motion. That makes sense, as a whirling dervish would want to ignore the motion-blurred world around them. One of the jobs of the precuneus is to form a perception of the position of the body in reference to the surrounding space. This makes sense as well, downplaying the need to constantly update estimates of body orientation during the prolonged rotation. These brain changes help explain how—with years of dedication and training—the Sufi whirling dervishes are able to use their mystic dance to approach enlightenment, rather than the toilet.

A Swiss team has been investigating the neuroscience behind the lulling effects of rocking.[14] Combining human and animal experiments, they have pieced together a compelling story. Using EEG (electroencephalography), the characterized brain waves of sleeping humans and mice, both with and without rocking. They found that rocking made it quicker to fall asleep, and improved sleep quality. With rocking, subjects didn't wake up as often during the night, spent more time in deep sleep, and had several patterns of brain activity thought to promote neural synchrony and help with memory consolidation. To test that idea further, they gave the humans a memorization test and found far better performance when they were tested after a night of rocking, versus a night of sleeping on a regular bed. It's pretty amazing, and all science needs to be repeated, but perhaps we'll see a market for adult rocker beds in the future! In the mouse experiment, the research team compared rocking-induced sleep benefits between normal mice, and a strain of mutant mouse who don't have a utricle. Recall that the utricle senses tilt and sway. In the utricle-deficient mice, none of the sleep benefits of rocking were present, leading the researchers to conclude that rocking works by stimulating the vestibular system. As goes the popular saying: "If the bed is a rockin', then you might be using vestibular stimulation to improve sleep quality and memory performance through utricular stimulation."

Patients with vestibular disease often complain of "brain fog." They report that they have trouble thinking clearly, with slower thoughts, difficulties remembering things, and trouble thinking abstractly. The brain fog clouds their thinking. Curious about this, I collaborated with a group of researchers at the Medical University of South Carolina, including Dr. Habib Rizk.[15] We gave patients with different vestibular diseases a standard questionnaire that asked about cognitive slips. It's called the Cognitive Failures Questionnaire, and it asks about the minor brain farts that each of us experience on a daily basis. "Do you read something and find you haven't been thinking about it and must read it again?" "Do you find you can't quite remember something although it's on the tip of your tongue?" We found that patients with vestibular disease did significantly worse than control subjects. Why is that? Why would vestibular disease affect not just spatial thinking, but thinking in general?

There are several theories as to why vestibular impairment could cause a slowdown in cognition. One widely circulated idea argues that since balance is normally a subconscious process, if it becomes conscious due to vestibular derangement, that would siphon cognitive resources that are normally used for general thinking. A second theory highlights some of the findings in this chapter, that vestibular circuitry is connected to areas in the brain like the hippocampus, which are used for spatial thinking. A third theory suggests that since vestibular disease is closely associated with some forms of psychiatric disease, like depression or anxiety, perhaps they are responsible for the impaired cognitive performance. However, I'd like to suggest a fourth theory.

In her book *Mind in Motion*, Barbara Tversky asks a profound question: "What is the nature of thought?"[16] How do we perform the mental calculations that allow us to form conclusions about the world around us? Implicitly, many of us assume that language is the substrate of thought. At first, this would seem to make sense. After all, isn't language how we express complex ideas? When I close my eyes, I can hear a little voice in my head. Doesn't that imply that language and thought are one and the same? However, this assumption seems to fall apart on several levels. First, animals clearly have some capacity for thought, and they don't possess language, at least not in the way that we conceptualize it. Consider that animals engage in social behavior, use tools, build homes, and teach their young. Second, many types of thinking, such as recognition, don't seem to involve language. If I asked you to describe your face, you would likely struggle, because your knowledge in that case is based on brain-facial recognition software, not on language. I don't get confused when I see other "middle aged, bearded men with piercing eyes, and graying, unruly hair," because that's not how that information is stored. Finally, as Harvard linguist Steven Pinker points out: if language was the stuff of the thought, then we would not be able to learn language as toddlers (how could we, if we couldn't think because we had no words yet), nor would we be able to come up with new words, as our mental boundaries would be set by our lexicon. Perhaps language isn't the essence of thought, instead it's an evolved methodology for thought communication.

Professor Tversky suggests an alternate answer: the foundation of thought is motion and space. We originally developed the ability for spatial thinking in order to navigate our environment, finding food and avoiding predators. However, that very same neural architecture that creates cognitive maps of the environment can also be used to create cognitive maps of ideas. These maps allow us to evaluate ideas by conceptualizing them as a network, with different relationships encoded by their position and connections within space. Related ideas may be *closely* linked, unrelated ideas may be *worlds apart*, and even time can be represented by order (my parents came *before* me).

Over eons, humans wandered the African Savannah, creating mental maps of the world. We localized resources and dangers: the grove of berries, the clean watering hole, the dry cave, the dangerous predators. However, unlike other animals, we had a large brain built for language and problem-solving. It's intriguing to wonder if our brains place ideas into this neural substrate, using virtual space as a medium for the processing and storage of ideas and concepts.

Time for a break. You made it through the section on how the vestibular system is supposed to work. Take a stretch and eat some delicious berries. When you come back, we are going to learn about several illustrative vestibular diseases. We are going to find out what happens when the vestibular system doesn't work.

PART 3

WHEN THINGS GO WRONG

8

Like a Rolling Stone

Benign Paroxysmal Positional Vertigo (BPPV)

t's been said that half of medical school is necessary to learn the art of healing, and the other half is required to master the vocabulary. For me, it's a love-hate relationship. Some of the terms seem to be overly long and complicated for no reason at all. For example, the general name for my field of medical expertise is "Otolaryngology" (which is synonymous with Ear, Nose, and Throat). The name is too long, too hard to pronounce, too hard to spell and generally confusing for people. Furthermore, despite its length, the term isn't accurate. "Oto" translates to "ear," and "laryngo" translates to throat. That leaves out *most* of the disease areas under our umbrella, including the nose/sinus, face, neck, and mouth.

However, medical terminology can also reflect the storied history of humanity's struggle to understand and heal itself. Names of body parts, diseases, and treatments often reflect the heroes (e.g., Prosper Ménière) and victims (e.g., Lou Gehrig) of that epic battle. They can also reflect mythology: cherubism causes cheeks to swell, resembling Renaissance depictions of the childlike cherubs. The uppermost vertebra is called the atlas, in homage to Atlas, the Titan of Greek lore who carried the weight of the world on his shoulders. Names can recall past culture: the Fregoli delusion, a belief that different people in your life are actually the same person in disguise, is named after a late nineteenth-century Italian actor, renowned for his ability to quickly change roles. As a surgical resident, I once cared for a patient diagnosed with "Ondine's curse." While we generally shy away from describing a disease as a curse, this one is so particularly vicious, perhaps the moniker is warranted. The poor souls with Ondine's curse have a faulty drive to breathe. While awake, there are no issues. But breathing stops altogether once they fall asleep. Consider living a life where an afternoon nap could be fatal. As you would suspect, most do not survive their first night of life as a baby. But, with good fortune and astute physicians, some can survive, spending their time asleep connected to a mechanical breathing apparatus. In novels and plays based on old European folklore, Ondine is a spurned water sprite who cursed her former lover, causing his death when he forgot to breathe during a fateful kiss.

I would place Benign Paroxysmal Positional Vertigo (BPPV) in that first category. The name is unwieldy, causing many to refer to it as "loose crystals." The title does make sense when broken down: benign (not an immediate threat to life), paroxysmal (occurring in fits or spurts), positional (provoked by certain positions, like lying down), and vertigo (the subjective sensation of movement). Throughout this chapter, we'll refer to it as BPPV.

Understanding BPPV is critical. It's widely considered to be the most common cause of vertigo. In fact, it's so common that it's sometimes used mistakenly as a synonym for vertigo. According to prevailing usage, vertigo is a symptom, whereas BPPV is a disease, just one possible cause of vertigo. In a frequently quoted study, Doctor John Oghalai and colleagues found that 9 percent of older adults have undiagnosed BPPV.[1] It's not just vertigo; multiple studies have found that those with BPPV are more likely to fall.[2] While falling may seem trivial to younger folk, it's a leading cause of death and disability in older people. The good news is that the fall risk drops back to normal after the BPPV is treated. And, as we'll see, it's relatively easy to treat in most cases. In fact, I frequently challenge my trainees to think of another disease that can be cured in two minutes. So far, no one has come up with anything.

PARTICLE PHYSICS

In the chapter 3, we painted a portrait of the detailed workings of the inner ear. BPPV can be thought of as a logical but unintended consequence of that design. Dense crystals are required for the otolith organs to sense gravity as they settle downward. Semicircular canals sense head turns. What happens if the crystals break off from their usual location, in the gravity-sensing zone, and accidentally end up in the semicircular canals, in the head turn-sensing zone? It's not a theoretical question, as various diseases (and aging itself) can loosen the otoconia from their crystalline fortress. This unmoored flotsam will wander inside the maze of the inner ear, eventually settling downward, like river silt. If the wayward crystals settle in the semicircular canals, BPPV results. Certain head turns result in crystals moving around near the cupula, suspended in the endolymph fluid. The crystals drag fluid with them, causing currents, like a spreading ripple from a stone thrown into a pond. The cupula has no way of knowing what is pushing or pulling on it. As a rudder within the torus of the semicircular canals, natural head movements cause it to deflect. But anything else that also moves the cupula will be interpreted by the vestibular system as being caused by the natural movement that would cause the same degree of cupular movement. The system can be fooled into thinking that you are moving when you are not. And with loose crystals in the semicircular canals, it thinks you are moving fast. It has no idea that actually you have not pursued a second career as an acrobat for Cirque Du Soleil. You aren't spinning around the stage in a human-sized hamster wheel, but the inner ear doesn't know that. That is the key to understanding the symptoms of vertigo: the vestibular system is sending strong signals that you are moving, but in fact you aren't. But, because the brain trusts the vestibular system, you feel like you are moving. This jarring mismatch, where the brain is being told one thing by your eyes, and another by your ears, underlies the nausea that frequently accompanies vertigo.

With three semicircular canals, you'd think that each would be equally affected by BPPV. 1/3 superior canal, 1/3 horizontal canal, and 1/3 posterior canal—right? What's that? You saw through my misdirect, and suspect that one canal gets affected much more than the others? Well, clearly there's no fooling you. Approximately 95 percent of cases involve the posterior canal. Superior canal BPPV is so rare, most specialists have never seen a case. The culprit here is geometry. Whether sitting, standing, or lying down, the posterior canal sits in the most gravity-dependent portion of the inner ear. It's like the U-bend under a sink. Because it has a loop that dips downward, debris tends to accumulate there. Since posterior canal BPPV is so much more common than horizontal canal BPPV or superior canal BPPV, it's assumed that BPPV is posterior canal BPPV unless otherwise stated.

Posterior canal BPPV is triggered by rolling over in bed or looking up. During those movements, trapped crystals float away from the cupula, strongly activating the hair cells. The brain mistakenly believes that you are executing a series of rapid backflips. To compensate, the eyes rotate downward. This triggers a protective fail-safe, and the eyes are quickly reset to neutral position. Caught in a cycle of these two opposing forces, like a seesaw, the eyes rapidly flicker back and forth. This flickering of the eyes is nystagmus, and the vector of nystagmus permits rapid and accurate diagnosis by an observer. This is because of J. Richard Ewald's first law, that the evoked nystagmus will be in the plane of the activated canal. The converse of that law is that by observing the direction of the nystagmus, you can deduce the canal at fault. The vertigo of BPPV is quick and intense, usually lasting under a minute. However, it's quite common for those afflicted to feel woozy or off-balance in between episodes.

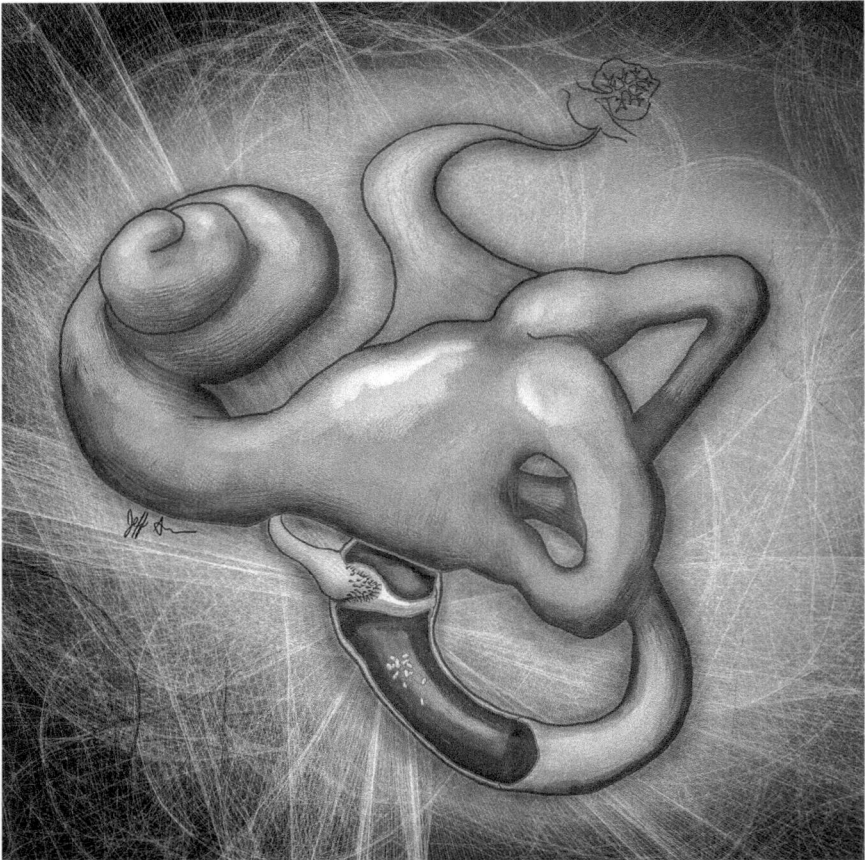

FIGURE 8.1 Benign paroxysmal positional vertigo (BPPV).

The curative treatment for BPPV was discovered relatively recently, in the early 1980s, by two physicians working independently: John Epley in Portland, and Alain Semont, in Paris. Today, both the Epley maneuver and the Semont maneuver are widely used throughout the world, and there is a mountain of clinical evidence attesting to their safety and efficacy. In the United States, Epley's treatment is far more popular, and is more of a household name. The story behind the discovery, though, is not what you'd expect, and is a cautionary tale in the power of medical orthodoxy.

In his excellent book, *Vertigo: Five Physician Scientists and the Quest for a Cure*, Dr. Robert W. Baloh traces the first written description of BPPV to 1921. The author was none other than Robert Bárány, who we met in chapter 1, famous for his Nobel Prize-winning work on the caloric vestibular response. He was born in Vienna, where he attended medical school. While there, he was inspired by a notable faculty member—Sigmund Freud—and asked him for an apprenticeship. He was rejected, because Freud found him "too abnormal," and therefore followed the advice of a friend—Gustav Alexander—to pursue a career in vestibular disorders. Alexander is known for his eponymous law, which describes a curious feature of the nystagmus due to a sudden one-sided vestibular dysfunction (for the curious, the nystagmus beats faster when you look away from the diseased ear then when you look toward it. It's thought to be related to the interaction of the vestibular eye movement signal, and elastic forces that always pull the eye toward a center position. When looking to the healthy ear, the forces are additive, so the nystagmus is faster, when looking toward the diseased ear, the forces are opposing, so the nystagmus is slower). Sadly, Alexander was later murdered by a former patient, who claimed that an operation to fix a collapsed nose had left his face deformed.

During World War One, Bárány served the Austro-Hungarian Empire as a field surgeon. He treated injured soldiers until his fortress was captured by the Russians in the spring of 1915, after a prolonged siege. Bárány was taken to Merv, a desert city in modern day Turkmenistan. While there, he wrote a letter to Harvey Cushing, the famed neurosurgeon, complaining about the mosquitoes and scorpions. Bárány received notification of his Nobel Prize while a prisoner of war. Not only that, but with the prize, Prince Carl of Sweden, where the Nobel Prize is based, petitioned Russia for Bárány's release, which was eventually granted. Therefore, in a certain sense, Bárány was granted his freedom because of his pioneering research. The award was tarnished with controversy, however, with accusations that he had not adequately credited others who had significantly contributed to his work. Things heated up with a formal inquisition from the University of Vienna, and Bárány left Austria for Uppsala, Sweden in 1917.

In 1921, Bárány published the first known report of BPPV. He described a young woman, who developed severe vertigo with nystagmus when she lay on her right side. The attack lasted thirty seconds. His observations on the pattern of nystagmus were quite accurate, noting, "a strong rotary (torsional) nystagmus, to the right with a vertical component upwards, which when looking to the right was purely rotary, and when looking to the left was purely vertical."[3] This is a nice validation of Ewald's first law governing vestibular function, which states that evoked eye movements will be in the plane of the canal. The plane of the canal is fixed within the head. For the posterior canal it's a diagonal vertical plane. The plane of the eyes within the head is not fixed, so as they move they come in and out of alignment with the canals. For the posterior canal on the right, the eyes will be in alignment when looking left. Therefore, as Ewald noted, the nystagmus in that position is purely vertical. It's like a pulley, with the posterior canal tugging on the eyeballs. When both face the same direction, the eyes get pulled up and down. But, when the eyes are turned to the side, the pulley is still pulling them with the same direction of force. Instead of going up and down, they rotate. Bárány mistakenly blamed the otolith organs for causing the nystagmus, rather than the semicircular canals.

Whereas the Epley and Semont maneuvers treat BPPV, the Dix-Hallpike maneuver is used to diagnose it. Introduced in 1952, it's still widely used today. Charles Skinner Hallpike was a prodigious researcher and influential ear surgeon at Queen Square, London. Margaret Dix was his junior coworker. Both had their own medical problems, Hallpike suffered from Legg-Perthes disease, which resulted in hip stiffening, and Dix had been injured in a World War Two bombing raid, losing vision in one eye. In an article published in the "Proceedings of the Royal Society of Medicine," Dix and Hallpike describe their strategy for approaching "this strange and dramatic disorder."[4] They described turning the patients head thirty to forty-five degrees to one side, and then "briskly" lying them back, with the head below the body by thirty degrees. They also call attention to the typical delay between positioning and the onset of the nystagmus—up to six seconds. There is even a suggested blueprint for a large mechanical contraption to twirl patients into the correct position.

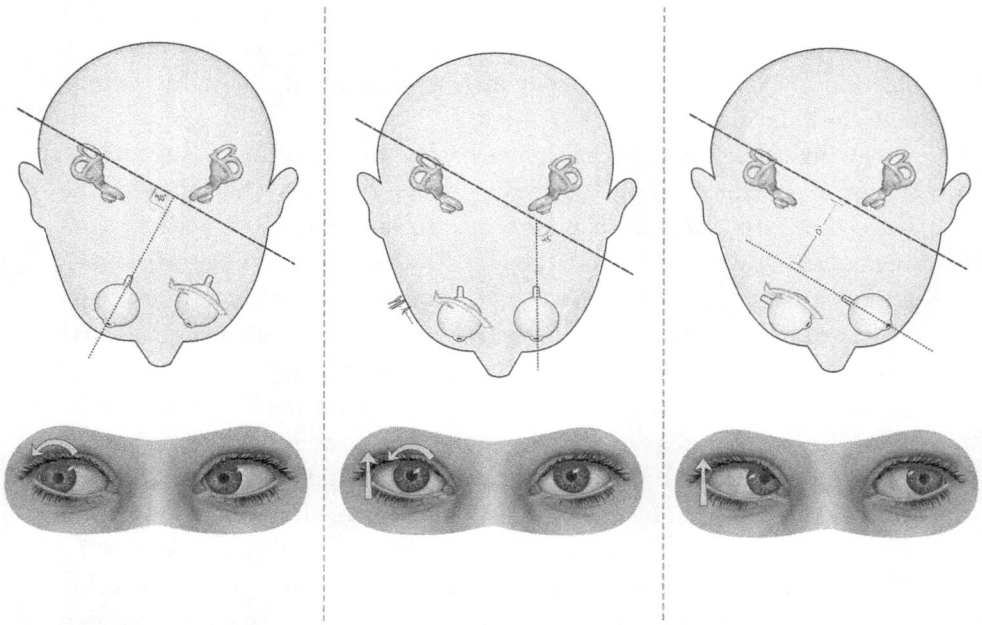

FIGURE 8.2 BPPV nystagmus depends on eye position.

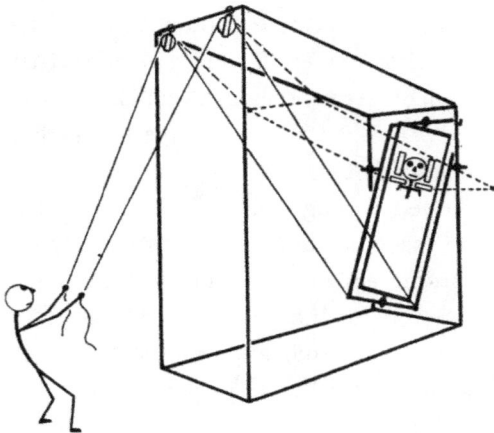

FIGURE 8.3 The Dix-Hallpike apparatus. From M. R. Dix and C. S. Hallpike, "The Pathology, Symptomatology and Diagnosis of Certain Common Disorders of the Vestibular System," *Annals of Otology, Rhinology, and Laryngology* 61, no. 4 (December 1952): 987–1016.

Dix and Hallpike correctly assume that this variant of vertigo is caused by the inner ear, and specifically, the "undermost" ear (while in position). Like Bárány, however, they erroneously assumed that the otolith organs were the culprit.

The inner ear was soon confirmed beyond doubt as the cause of BPPV, as several patients were cured after surgical destruction. This intrigued the young Dr. Harold Schuknecht. He had seen several patients with a recurring pattern of symptoms. First, they would get severe spinning vertigo, lasting a few days. Later, they developed positional nystagmus. By studying postmortem temporal bones, he found several anatomic clues. In those cases, the upper part of the inner ear, containing the utricle, superior canal, and horizontal canal, would be severely degraded from inflammation. The lower part of the inner ear, containing the saccule and the posterior canal, would appear normal. Using a rationale that would have made Sherlock Holmes proud, he deduced as follows: (1) only healthy organs can send dynamic signals to the brain, so the posterior canal or the saccule must be the cause; (2) the saccule is excluded, since it was not thought to be capable of generating vertigo; (3) the true identity of the villain is therefore revealed as the posterior canal. He went further, reasoning that the crystals of the utricle may have been scattered during the damage and settled on the inner side of the cupula (the side facing the vestibule, not the side facing the posterior canal). The theory was termed "cupulolithiasis"—i.e., stones in the cupula, in an influential 1969 paper.[5] It should be noted that this particular pattern of inner ear damage is only seen in a small fraction of cases of BPPV, which today we would call BPPV following superior vestibular neuritis. As we've seen, there are actually two vestibular nerves, so sometimes just the top one becomes inflamed, damaging its connected organs, but leaving the bottom organs—the saccule and the posterior canal—unharmed. Schuknecht rose to prominence and was pronounced chief of the illustrious Massachusetts Eye and Ear Infirmary. Due to his stature, the "cupulolithiasis" theory was widely believed, despite several flaws, including the fact that it predicted eye movement in the opposite direction of what was observed.

In the span of fifty years, the magnifying glass of science had zoomed in, first on the inner ear, and after some wrong turns, finally on the posterior canal. Dr. Richard Gacek was intrigued and developed an intricate operation to sever the nerve to the posterior canal to cure BPPV. That nerve, a branch of the larger inferior vestibular nerve, is miniscule, like a strand of hair. My former chairman, Dr. Rick Chole, once told me, "The operation was wonderful, and it worked. The only problem was that he was the only one who could actually do it."

THE CRYSTAL SKULL

In the early 1980s, Dr. John Epley of Portland, Oregon put the last piece of the puzzle in place. He envisioned the inner ear, and correctly deduced that free-floating crystals were in the posterior canal. With this insight, he began to experiment with different repositioning treatments, to move the crystals away from the posterior canal. The maze of the inner ear became the maze of childhood puzzles, and a series of precise movements were required to navigate marbles to the goal. Together with Dominic Hughes, an audiologist, he built a model of the inner ear, which helped them picture the sequence of movements necessary to relocate the crystals. It worked like a charm. Dr. Epley submitted his groundbreaking work to the *American Journal of Otology* and was rejected. "Findings were not consistent with existing theory." He tried two other journals, to no avail.

Dr. Epley continued to push his new treatment. But the world pushed back. As relayed in the *Oregonian*, Epley continued to experience heavy skepticism throughout the 1980s.[6] In one instance, he performed his treatment under anesthesia, for a patient who was wheelchair bound with intractable symptoms. The anesthesiologist thought he was incompetent and filed a complaint with the hospital. Another time, a disgruntled physician rudely walked out of his lecture. Semont published his work in 1988, and after ten years of struggling, Epley's work was finally published in 1992.[7] However, his troubles did not end there. In 1996, the Oregon Board of Medical Examiners initiated an investigation into unprofessional conduct. The case was adjudicated in Epley's favor, but years of skepticism and ridicule had left their toll.

In 1990, two surgeons in Canada—Lorne Parnes and Joe McClure—published a paper on a new surgical technique for BPPV.[8] Instead of demolishing the whole inner ear, or attempting Gacek's tricky operation, they proposed that the posterior canal could be accessed through the mastoid bone, and the interior of the canal occluded. With the insides blocked with tissue, no endolymph movement would be possible, and therefore aberrant crystals would not be able to fool the gullible cupula. The technique was safe and reliable and is still in use today for the rare case where multiple rounds of Epley's and Semont's treatments are not effective. A few years later, during surgery, they were able to see "debris" floating inside the delicate membranous tunnel inside the canal. During two of their surgeries, the debris was removed, and sent to one of my former professors, Dr. Rick Chole. He recounts that the material sat in cold storage for quite some time, but it was finally analyzed and published in 2016. Using scanning electron microscopy, the microscopic makeup of the debris was clearly seen.[9] As had been suspected since Schuknecht's work in the 1950s, there were indeed wayward crystals inside the posterior canal. In a corroborative study, Hui Xu and colleagues found that the source of the crystals—the utricle—does not work properly in the majority of cases of BPPV, presumably due to depletion of its crystal lattice.[10]

I have to believe in the process of science. In the saga of BPPV, personalities, world events, dogma, and chance all played a part. However, the candle of knowledge was carried forward, passed between generations. Not omniscient, but self-correcting, forward progress was assured by the scientific process, despite false passages. Hypotheses were tested and retested, until only truth remained. As we'll see in the next chapter, on Ménière's disease, this process doesn't always yield timely results, and that disease has no cause, no cure. But I remain hopeful that in time, all the secrets of the ear shall be illuminated.

9

Unsolved Mysteries

Vestibular Migraine and Ménière's Disease

L ife is full of mysteries. Are we alone in the universe? How are members of our species simultaneously capable of writing the enduringly beautiful music of the Beatles, and murdering John Lennon? Why is the speed of light a constant value, no matter the speed of the observer? Do other people think my butt looks big? Is there life after death? Does my cat even like me? When will flying cars be available for lease? Did men's fashion stop evolving after we discovered the suit and tie? The list goes on, and it can seem overwhelming.

Yet, we can take comfort in the great questions our collective mind has answered. The sun, ninety-three million miles away, with 99.9 percent of all the mass in our solar system, generates enough gravitational force to smash atoms, fusing them in a raging thermonuclear furnace that will burn bright for another five billion years. The human genetic code is composed of three billion base pairs of DNA, sufficient software to direct a single cell to self-assemble into a fully functional human child. New Jersey is not as bad as everyone says, once you get off the turnpike.

At the time of writing, the cause of both Ménière's disease and vestibular migraine remain mysterious. Both are diagnosed based on clinical criteria—checklists of symptoms—and not on diagnostic tests. Both are common causes of repeated episodes of vertigo, but as we'll see, vestibular migraine is far more common. And despite not knowing the cause, both do have effective, albeit imperfect treatments. However, the real reason they are lumped together in the same chapter is because they are related, like two sides of a coin, or the two faces of Janus. Consider that half of all patients with Ménière's disease *also* have had migraines in the past. It seems likely to me that until we understand how migraine affects the inner ear, we won't understand its sister, Ménière's disease.

BLACK, BILIOUS MATTER

Before we dive in, it's worth discussing migraine and vestibular migraine, as there are several common myths out there. Migraine is a neurologic disease that produces many different symptoms, the most common of which is headache. It's important to know that migraine can produce many other symptoms as well, frequently without a headache. For example, migraine can cause someone to see visual mirages, like shimmering shapes, iridescent sparkles, and enlarging blind spots. These ocular migraines can occur with a headache, or without. Some patients with migraine get abdominal pain, some get temporary paralysis, and as we'll see, some get dizzy. So, migraine and headache are not synonymous terms.

Funnily enough, about five years after dedicating my research energies to studying migraine and dizziness, I experienced a few migraine auras. The first time I was in clinic, moderately peeved as I was responding to a mountain of frustrating emails just before the first patient of the day, when the light show began. For fourteen minutes, I saw shimmers and sparkles float diagonally across my field of view. It looked to me like the wavy distortions you see in the ophthalmologist's office, during visual field testing. Fortunately, no headache followed. My second aura was a few weeks later. I was at a conference and had just started an hour-long talk on my research into migraine and dizziness when I was treated to private fireworks in my head. I do enjoy public speaking, so I wasn't stressed out at the time, but perhaps the universe couldn't resist the grand irony of the situation.

Vestibular migraine is a type of migraine attack, where the primary symptom is dizziness, and not a headache. Sometimes there is an associated headache, and sometimes there isn't. However, there is always a clue that migraine is the culprit—either a prior history of migraine headaches, or typical "migrainous" symptoms occurring during the dizziness—like sensitivity to light or sound. In my clinic, vestibular migraine is the most common diagnosis among dizzy patients. Despite how common it is, it's also by far the most overlooked cause of dizziness. In fact, many of my own patients are surprised to be diagnosed with vestibular migraine and have previously been told that their dizziness or vertigo had other causes.

Sometimes a diagnosis can be hard to make because someone has an incredibly rare condition. Sometimes a diagnosis can be hard to make because someone has an incredibly common condition, so common it's overlooked. Migraine is remarkably common, affecting one in ten people worldwide. The World Health Organization puts migraine as the third most common disease in the world, following anemia and hearing loss. It's the most common neurologic disease, and the most common cause of disability among working-age people. In a memorable 2005 graduation speech, David Foster Wallace recounted a parable of two fish, who encounter a third. He says, "Morning boys, how's the water?" As he swims away, the other two fish look at each other. One asks, "What the hell is water?" For the fish, water is everywhere, too ubiquitous to be appreciated. In the realm of vestibular disease, it often feels like migraine is water, right in front of our faces, but we see right through it.

In addition to headache and dizziness, migraine can cause many of the most common symptoms in my field of medicine: otolaryngology (ear, nose, and throat). This includes neck pain, sinus pressure, watery eyes, eye redness, nasal congestion and drainage, swelling or flushing of the face, and ear pressure. Therefore, I advise my trainees to keep migraine high up on the "differential diagnosis," which is the mental list of possible causes for any set of symptoms.

Vestibular migraine results in an assortment of secondary problems. In one study, we used a large national database to investigate three specific issues. We found that compared to the general public, those with vestibular migraine were seven times more likely to have trouble thinking, six times more likely to have decreased mobility, and two-and-a-half times more likely to have a fall. They also missed, on average, seven more workdays per year than those without vestibular migraine.[1] The ratios held true, even when we controlled for confounding variables within our analysis. Unfortunately, I don't think the results of this study would surprise anyone who routinely cares for these patients.

Prosper Ménière first described his eponymous disease over 160 years ago. Vestibular migraine, on the other hand, has only received modern attention over the last thirty years. However, written descriptions appear far earlier. The following account is from Aretaeus of Cappadocia, written in the first century CE:

> If darkness possess the eyes, and if the head be whirled round with dizziness, and the ears ring as from the sound of rivers rolling along with a great noise, or like the wind when roars among the sails, or like the clang of pipes or reeds, or like the rattling of a carriage, we call the affection Scotoma (or Vertigo); a bad complaint indeed, if a symptom of the head, but bad likewise if the sequela of cephalæa, or whether it arises of itself as a chronic disease. For, if these symptoms do not pass off, but the vertigo persists, or if, in course of time, from the want of any one to remedy, it is completed in its own peculiar symptoms, the affection vertigo is formed, from a humid and cold cause. But if it turns to an incurable condition, it proves the commencement of other affections—of mania, melancholy, or epilepsy, the symptoms peculiar to each being superadded. But the mode of vertigo is, heaviness of the head, sparkles of light in the eyes along with much darkness, ignorance of themselves and of those around; and, if the disease go on increasing, the limbs sink below them, and they crawl on the ground; there is nausea and vomitings of phlegm, or of yellow or black bilious matter. When connected with yellow bile, mania is formed; when with black, melancholy; when with phlegm, epilepsy; for it is liable to conversion into all these diseases.[2]

While there's always a murkiness to interpreting ancient writings through modern eyes, note that Aretaeus links dizziness, tinnitus, imbalance, aura, nausea, and headache—all of which can be a part of vestibular migraine. The descriptions of tinnitus are noteworthy as well, poetically drawn from everyday experiences, 2,000 years ago. A modern writer might say regarding tinnitus, . . . and the ears ring as from the sound of radio static, played at high volume, or like the roaring of a jet engine, a great noise, or the hum of an air conditioner, or like the chime of an email alert . . .

Despite Aretaeus's description, vestibular migraine wasn't appreciated until the early 2000s. My program director in residency was Joel Goebel, a world-famous vestibular doctor and educator. He had published some of the earliest studies of vestibular migraine, so I asked him what the landscape of vestibular migraine looked like in the 1990s. Dr. Goebel laughed. "It was a prairie. There was no landscape. It was just a prairie." He continued, "The neurologists didn't recognize the issue, the otolaryngologists still wanted to squeeze dizzy patients and vertigo patients into the labyrinth, or, if they were chronically dizzy, they squeezed them into psychogenic vertigo. You're hyperventilating. Your brain is overactive. Learn to calm down. Take a sedative. It just wasn't known." Stuck in the prairie, Goebel recounted that patients "bounced around. They suffered. . . . Migraine was the gorilla in the room." But no one was paying attention to the gorilla.

Dr. Goebel ran his clinic like a vestibular detective. He would collect all the clues, and then solve the mystery, with a grand reveal at the end of the visit. I was curious about what drew him to the vestibular system in the first place. He told me it was, "The absolute intricacy of the system. It's almost mysterious. It's always running, always on, and yet we're not consciously aware of what it's doing." Dr. Goebel wasn't just an expert in vestibular medicine. He was also an expert at connecting with people. During our graduation, he paraphrased Maya Angelou: "Years from now, your patients are not going to remember what you said. They aren't even going to remember what you did. But all of them are going to remember how you made them feel."

Why does vestibular migraine cause dizziness? Standard vestibular test batteries are usually normal, indicating that vestibular information being transmitted to the brain is reliable. Why then do sufferers experience dizziness, vertigo, disequilibrium, and a general sense that their body isn't properly oriented with respect to their environment? An emerging body of evidence suggests that the issue may lie with the brain's ability to integrate multisensory information (visual, vestibular, and somatosensory) into a single, cohesive mental representation of one's position within space.

A team of researchers from Johns Hopkins approached this question by investigating our natural ability to discern verticality.[3] This is accomplished in a dark room, where seated subjects are asked to judge whether a red line on a display monitor was slanted to the left or right of true vertical. The red line, like a hand on a clock, could appear at any degree of rotation (so twelve o'clock would be true vertical). With repeated measurements over a hundred trials, investigators formed a detailed assessment of each subject's ability to judge true uprightness. The measure on Earth of whether something is truly vertical is if it aligns perfectly with the pull of gravity. So, this is a test of one's ability to sense gravity. With heads held straight, those with vestibular migraine performed the same as normal control subjects: both groups were really good at aligning the red line with true vertical, to within a degree or two of accuracy. However, when asked to do the same task while their heads were tilted to one side, those with vestibular migraine performed more poorly. This task is harder, because it involves integration of multiple streams of sensory information, including visual, neck position sensation, and vestibular. When your head tilts to the right, your eyes roll a little to the left to compensate. This is called an ocular counter-roll, and it's a vestibular reflex. However, unlike other vestibular reflexes, the ratio of head movement to eye movement isn't one to one. Instead, the ratio is closer to five to one, such that if your head gets tilted twenty degrees to the right, your eyes will roll to the left about four degrees to compensate, which rotates the entire visual world. In order to accurately assess verticality, you need to consider the vector of gravity according to the vestibular system and then adjust it for the degree of head tilt and eye roll. Hence, multisensory integration is required. For rightward head tilts, which by chance happened to coincide with most subject's altered perception (i.e., during day-to-day life, they felt the world was tilted to the right), subjects with vestibular migraines performed significantly worse than normal subjects, and on average were off about three degrees.

The research team, led by Dr. Amir Kheradmand, had previously tried to figure out what part of the brain is responsible for the multisensory, integrated perception of gravity. They had a hunch that an area called the temporoparietal junction was involved, based on functional imaging studies. To figure this out, the research team used a technique that sounds like it's straight out of science fiction. Transcranial magnetic stimulation (TMS) works by using magnets to induce electric currents in targeted brain regions. The technique is noninvasive, using magnet coils outside of the head to cause brain currents, and side effects are rare. The research team had volunteers estimate verticality, using a protocol similar to their previously mentioned study. The volunteers were tested under several conditions, including with their heads held upright, or tilted to the right or left, with the TMS either on, off, or stimulating a different brain area not thought to be involved in gravity perception. They found that with head tilts, and TMS stimulation of the temporoparietal junction, they were able to consistently change volunteer's perception of Earth vertical. This did not occur when the TMS system was off, or when it stimulated other brain areas. Therefore, it's fair to say that this particular region of the brain—the temporoparietal junction—plays an important role in combining different streams of sensory information into an estimate of true verticality, and it may be dysfunctional in those with vestibular migraine.

There are a few other theories trying to explain how migraine can cause dizziness. One theory supposes that vertigo is simply a migraine aura. On its face, this makes sense. If a visual aura causes visual hallucination (seeing things that aren't there), then a vestibular aura should cause a vestibular hallucination (feeling movement that isn't there). Feeling movement of yourself or the world that isn't occurring is the precise definition of vertigo. We know that a visual aura occurs as a wave of electricity washes over the occipital lobe, which is the visual processing center of the brain. It's called "cortical spreading depression," because it involves the surface of the brain (cortex), spreads, and depolarizes neurons (depression). Logically, if this neural storm landed in a vestibular center (like the parieto-insular cortex), then vertigo should ensue.

Recent years have seen a boom in migraine medications, and most new drugs are based around one particular brain chemical. CGRP (Calcitonin Gene Related Peptide—the name is actually not important, so it's easier to call it by its acronym) is part of the neuroinflammatory process. Numerous studies have shown that if you block CGRP release, you can block migraines. Many of my own patients have described CGRP-blocking drugs as a "game changer." In the opposite experiment, in susceptible individuals, you can trigger a migraine by injecting CGRP into the bloodstream. You may wonder who in the world would sign up for an experiment that causes intense pain to the participants. It's a testament to how disabling migraines can be, to the point where sufferers are willing to do anything to advance the science.

There are twelve main nerves in the head, called the cranial nerves. They are named from front/top to the back/bottom. Medical students are taught to remember them using mnemonics, and I sincerely hope the memory aids become less vulgar over time. Fear not, dear reader, I shall spare your delicate ears from the bawdiness. The fifth nerve, the largest, is called the trigeminal nerve. As fans of Latin will recognize, the name translates to "triplet," corresponding to the three major branches of the nerves. It's mostly a sensory nerve, plugged into the eyes and forehead (first branch), the cheek area (second branch), and jaw area and tongue (third branch).

The trigeminal nerve seems to go haywire during a migraine, releasing CGRP, and sometimes causing face numbness, face tingling, or sinus pressure. Electrically tickling the trigeminal nerve can cause changes in the inner ear, including leaky blood vessels, and as we'll see this may provide the experimental basis for the link between migraine and Ménière's disease, whose hallmark is swelling within the inner ear.[4] Furthermore, CGRP is found within the inner ear, and in animal models it has effects on sound sensitivity and vestibular reflexes. When mice are genetically engineered without CGRP, they have poor balance.[5] In one experiment, the CGRP-deficient mice keep falling off a rotating balance beam, unlike normal mice. So, it seems that CGRP may be part of the reason that those with vestibular migraine suffer dizziness.

To test that theory, I ran a clinical trial. In the study, patients with vestibular migraine were randomly assigned to either get a CGRP-blocking drug or a placebo for three months. The drug in the study was called galcanezumab, and it's sold under the trade name Emgality. Full disclosure: the study was paid for by Eli Lilly, who makes the drug. However, I approached them to do the study, not the other way around, and they had no control over the data or the analysis. They just provided the drug, the placebo, and the financial support necessary to do this type of work. Importantly, the drug and the placebo looked identical and were packaged the same, so the study team and the patients didn't know what they were getting. That is called blinding, and when both patient and physician are blinded, it's called double blinding. Now, while this may seem maddening, a randomized, double-blind, placebo-controlled clinical trial like this one is the best way to figure out if a drug actually works. Part of the reason is that many people will have a strong placebo response, meaning that they will get better even if they don't get the treatment. In addition, diseases like vestibular migraine naturally fluctuate over time, with spells of severe symptoms, and periods of tranquility. Therefore, the goal of the clinical trial is to find out if a treatment has additional benefit over the expected placebo response. And, since people naturally vary with their disease severity and also with a million other factors, the hope is to equal out all the unknowns through the randomization process. That way, at the end of the trial, the only thing that was really different between the two groups was whether they got the drug or the placebo.

We called our study the INVESTMENT trial. That's because I was taught that all good clinical trials have to have a good acronym. INVESTMENT stood for INvestigating VESTibular Migraine Emgality treatmeNT. The trial ran into some problems, including COVID-19, and we were affected by supply chain issues. Therefore, we ended up with fewer patients than we were hoping for. Despite that, we were able to clearly show that blocking

CGRP was really helpful for vestibular migraine. In the baseline month, before any treatment, both those getting placebo and those getting active drug averaged eighteen dizzy days. That's a good thing, because it means the randomization worked, and both groups were similar before starting treatment. We found that, on average, those getting a placebo had 12.5 dizzy days in the final month of treatment. So, people improved by just getting a placebo. But those who got the treatment dropped from eighteen days in the baseline month to just 6.6 dizzy days in the final month. In other words, they had 5.7 fewer dizzy days in the final month of treatment than those who got the placebo.

We also measured treatment success in a couple other ways. We had patients fill out the Dizziness Handicap Inventory, which is a widely used questionnaire to assess dizziness severity. We also had patients fill out VM-PATHI, which is the Vestibular Migraine Patient Assessment Tool and Handicap Inventory. My team had developed and validated VM-PATHI a few years prior because we wanted a way to measure all the symptoms of vestibular migraine, and none existed. By comparing scores before and after treatment, we found that blocking CGRP reduced both the Dizziness Handicap Inventory and also the VM-PATHI score much better in the treatment group than the placebo group.

There are a lot of implications for the study. The most obvious, which shouldn't even have to be said, is that vestibular migraine is a *real* disease. It's not some made up thing in people's heads caused by anxiety or whatever. We can see that clearly because we didn't treat any anxiety in this study. The second thing, which also should probably not need to be said, is that vestibular migraine is a migraine variant. That is why a targeted migraine medication worked so well. The third implication is that CGRP release in the brain causes dizziness. I am planning some future studies to try to better understand why that is.

Many with migraine have sensory sensitivity. Normal sensations that wouldn't bother the average person become noxious. Lights become blinding. Sounds are deafening. Touch becomes painful. Smells are vile. Those with vestibular migraine can suffer all that, with an addition. They develop a vestibular sensitivity to motion. Movements, once routine, now cause unease. Quick head turns, rising quickly from sitting to standing, and head bobbing/shaking are all avoided. In addition, many of those with vestibular migraine are prone to seasickness and carsickness, likely through a similar mechanism.

Currently, just like migraine headache, there's no cure for vestibular migraine. Instead, we rely on treatment, and there are two main categories: Abortive/symptomatic medications are only taken during a migraine attack, to try to end the attack as quickly as possible and relieve symptoms. Prophylactic medications are taken every day (or for long-acting formations, present every day), and aim to decrease the frequency and severity of migraine attacks. For vestibular migraine, the most common acute medications tend to be anti-nausea agents, like ondansetron (Zofran), or vestibular suppressants, like meclizine (Antivert), or benzodiazepines (e.g., Valium or Klonopin). That way, the two most common symptoms—nausea and dizziness—can be targeted and hopefully eliminated.

Anyone who suffers from migraines is familiar with triptans. Medications like sumatriptan (Imitrex), rizatriptan (Maxalt), and zolmitriptan (Zomig) are widely used and immensely popular. Numerous studies have proven that triptans are helpful for migraine headaches, leading to a natural question: Are triptans helpful for vestibular migraine? The question is practical, because those suffering need relief yesterday, but also theoretical. Is vestibular migraine similar enough to migraine that the same drugs work, or is it distinct enough that the proven therapies for headache won't work for dizziness and vertigo? A research team from several notable institutions, including UCLA and the Mayo Clinic, set out to answer this question. Their results haven't been published yet, but they were presented the 2022 Bárány Society meeting. There, Dr. Jeff Staab explained the preliminary (and subject to change) findings: Rizatriptan did not reduce vertigo compared to a placebo in vestibular migraine, measured one hour after the onset of symptoms. However, at the twenty-four-hour mark, it did help with motion sickness and with dizziness.[6] To me, this implies that vestibular migraine are a little different than migraine headaches, and that we shouldn't just assume that because a medication works for migraine headaches, it will work for vestibular migraines.

A whole variety of prophylactic medications have been enlisted for vestibular migraine. These meds, borrowed from migraine headaches, are shockingly diverse. Some prevent seizures, some control blood pressure or heart rhythm, and others were originally formulated as antidepressants. However, they all share a key feature: when compared to a placebo in a head-to-head trial, they won. With little data to go on, vestibular doctors just use them, assuming that migraine headaches and migraine dizziness are similar enough for them to work. A 2015 Cochrane review, widely considered to be the definitive authority for evidence-based medicine, found no trials of sufficient quality on vestibular migraine.[7] Unfortunately, not much has improved since then, with rare exception. A German study called PROVEMIG was published in 2019.[8] The authors studied a popular migraine medication called metoprolol and compared its performance to a placebo. The study ended early to due recruitment problems. However, at its end the researchers had recruited 130 subjects, out of a planned 266. Analysis of the available data showed no difference between metoprolol and placebo, indicating that the drug was not effective for vestibular migraine.

There's a grand irony in using drugs to treat dizziness. Dizziness is listed as a side effect of almost every medication out there! I now warn patients of this twisted fact when prescribing, hoping to maintain my authority and their loyalty. Even worse, most patients with vestibular migraine are sensitive to medications. Quanta of light, innocuous to most, burn those with migraine, and snippets of sound, normally ignored, pierce the migraineur's ears. Similarly, dosages of medications that are commonly tolerated by most people can ruffle those with vestibular migraine. With bodily sensitivity dialed to maximum, they perceive the world differently. For those with migraine, a common rule of prescribing is "start low and go slow." Use the lowest available dosage and give the body plenty of time to acclimatize before increasing the dose. Despite that, for many patients, drug treatment simply isn't feasible.

Fortunately, non-pharmacologic treatment options do exist. The simplest is trigger avoidance. I've surveyed my patients, and the most common triggers for vestibular migraine, in descending order are: motion, stress, busy visual scenes, bright lights, scrolling on a screen, sleep disturbance, loud noises, dehydration, barometric pressure changes, air travel, skipped meals, hormonal changes, certain foods or beverages, allergies, and certain smells. Some can be avoided completely (e.g., skipping meals), some can be minimized (e.g., scrolling on a screen), while others can be hard to control (e.g., stress). Like many diseases, eating healthy and exercising regularly is helpful. Some vitamin supplements—magnesium and riboflavin—are routinely recommended for migraine headaches, it's unclear if they help vestibular migraine.

Recognizing that stress can be a major trigger for vestibular migraine, and also that half of those with migraine suffer from anxiety (for unknown reasons), we launched a study to see if treating that face of the disease would help overall. Mindfulness Based Stress Reduction (MBSR) involves formal instruction in the techniques and philosophies of mindfulness. Breathing exercises and yoga are incorporated as well. It's a contemporary take on the meditative practices of Eastern religions, trying to bridge ancient teachings with modern medicine. Participants are taught to experience the world and body sensations in a nonjudgmental fashion. Focus is placed on attention, intention, and attitude. At the Osher Center for Integrative Medicine at UCSF, classes also involve reading *Full Catastrophe Living* by Jon Kabat-Zinn, PhD. Kabat-Zinn is credited with bringing mindfulness into the medical sphere, and it's now clear that there is strong scientific evidence that mindfulness is helpful for a variety of medical conditions.

The philosophy of mindfulness reminds me of the poem "The Guest House" by Rumi, a Sufi mystic from the thirteenth century. Rumi was also the first whirling dervish, who we met in chapter 7.

> This being human is a guest house.
> Every morning a new arrival.
>
> A joy, a depression, a meanness,
> some momentary awareness comes
> as an unexpected visitor.
>
> Welcome and entertain them all!
> Even if they're a crowd of sorrows,
> who violently sweep your house
> empty of its furniture,
> still, treat each guest honorably.
> He may be clearing you out
> for some new delight.
>
> The dark thought, the shame, the malice,
> meet them at the door laughing,
> and invite them in.
>
> Be grateful for whoever comes,
> because each has been sent
> as a guide from beyond.[9]

In our study, subjects with vestibular migraine underwent an eight-week course in MBSR. We administered a set of questionnaires before and after the intervention. We also sent a daily text message to study participants, asking them to rate dizziness as none, mild, moderate, or severe. The study was funded by the Association for Migraine Disorders and allowed us to enroll twenty subjects. Not surprisingly, we found that anxiety and depression scores were improved. However, we were excited to find that dizziness had improved as well. It was still present in most cases, but not as severe. Analyzing the data, we found that the biggest change was due to a switch. Moderate dizzy days dropped, and mild dizzy days increased, their curves appearing as mirror images. It seemed that mindfulness was effective, by downgrading moderate dizziness to mild.[10] Based on my experience as a physician, I am a strong believer in the mind-body connection, and this study reinforced that.

We did find something unexpected during the experiment. Before and after the study, we administered a questionnaire called the "Cognitive Failures Questionnaire." It measures everyday brain slipups, like when you forget your keys, your shopping list, or the precise sequence of events that led to World War One. Surprisingly, scores significantly decreased (improved) during the study. There are quite a few interpretations. Perhaps without the constant distraction of dizziness, the mind was simply clearer and sharper. Or perhaps better scores were due to better moods, as it's well-known that those with anxiety and depression score worse on most tests. However, it's also possible that vestibular glitches directly degrade the capacity for abstract thought, and by treating the vestibular problem, we can also help dispel the brain fog. This gives me hope, as trouble thinking is a common complaint among my patients.

A major focus of my career has been to try to help those with vestibular migraine. We first tried to figure out how many Americans suffer from vestibular migraine and estimated that it affects about 3 percent of adults. We next developed and validated a questionnaire (aka a "patient reported outcome measure") to help quantify how bad someone's vestibular migraine symptoms are. We called it VM-PATHI (Vestibular Migraine Patient Assessment Tool and Handicap Inventory), to rhyme with "empathy." It's on our website, so that researchers and patients the world over can use it for free.[11] Finally, using our new tools, we are trying to study which treatments work, and which don't for vestibular migraine. When someone comes into my office, they just want to feel better. To feel normal again. With the help of colleagues, research team members, and my brave patients who volunteer for studies, we are trying to help all those who suffer from this common and terrible affliction.

TOO MANY YEARS OF MÉNIÈRE'S

I hesitate to write about Ménière's disease, for fear that we know too little. I suppose we should start with what we do know. There is clearly a degenerative disease that can occur, affecting just one of the inner ears. It causes terrible tinnitus, pounding pressure, violent vertigo, and shifting sensorineural hearing loss. In postmortem examinations, under the microscope, a particular swelling is visible within the innermost partition of the inner ear: this is called "endolymphatic hydrops." It's unclear if this swelling is the cause of Ménière's disease, an effect of Ménière's disease, or simply a byproduct. So, if you are a disciple of fact, then you can skip the rest of the chapter. Unfortunately, we must wander into the realm of conjecture to continue our conversation.

Because we fundamentally don't understand Ménière's disease, the most effective treatments are destructive in nature. Without the knowledge to cure, we instead demolish. It's a scorched-earth strategy, losing the battle to win the war. There are three methods of ear annihilation. With a "vestibular nerve section," a small window into the skull is created, and used to sever the vestibular nerve. Without a connection to the brain, the ear may continue to have Ménière's disease, but in silence, without symptoms. Secondly, with a "labyrinthectomy," the inner ear is surgically extirpated. A drill spinning at 80,000 RPM bores a tunnel through the honeycombed bone of the mastoid, then through the rock-hard capsule of the inner ear, accessing the delicate membranes and sensory organs of the vestibular apparatus, which are unceremoniously scooped away. Unlike cutting the nerve, the labyrinthectomy does sacrifice any remaining hearing, so it's only offered when hearing has degraded beyond salvage. Finally, it's also possible to chemically poison the inner ear. Recognizing that a certain group of antibiotics—aminoglycosides—cause tinnitus and hearing loss as a side effect, physicians began exploiting that finding in the 1970s. When aminoglycoside antibiotics are injected through the eardrum into the middle ear, some of the medicine is absorbed by the inner ear. Once inside, the antibiotic kills hair cells. There's a whole range of aminoglycosides, and some have much higher chemical attractions to vestibular hair cells than cochlear hair cells. That equates to a higher likelihood of causing the intended vestibular damage and avoiding the unintended collateral damage to hearing. Gentamicin is currently the poison of choice for this purpose.

In medical school, I recall being frustrated by an overwhelming number of treatment choices for some diseases. A wise, older physician advised me: "You see, when a disease has more than a few treatment options, it's because none of them work." In my defense, I was unduly influenced by TV doctors. The characters on *House, Scrubs,* and *Gray's Anatomy* spent most of each episode seeking the right diagnosis. They were detectives, using clues and deductive reasoning to solve the mystery illness. After the dramatic reveal—Chlamydia!—a quick cure was administered, which always worked, segueing into the closing scene of a cured patient leaving the hospital. I had assumed that diagnosis was the hard part. In reality, many of our treatments only help some of our patients, for some percentage of their symptoms, some of the time. Cures are frustratingly elusive.

Ménière's disease has a wide range of nondestructive treatments. They include low-salt diets, diuretics (water pills), betahistine, steroids (pills and injections), vasodilators, vasoconstrictors, antihistamines, anesthetics, sedatives, vitamins, antipsychotics, ear tubes, ear massagers, antivirals, and surgeries. Recall that the inner ear is protected by the strongest bone in the human body: the otic capsule. Once the capsule is entered, another problem is encountered. The delicate vestibular machinery is far too small for human surgeons to do anything other than destroy. Even with the aid of powerful (and expensive!) operating microscopes, and micro-instruments, we are like a child trying to repair a broken watch with clumsy fingers. Furthermore, in Ménière's disease, the swelling (hydrops) affects the endolymphatic space, which is inside the larger and surrounding perilymph space. However, there is one area where the endolymph space is surgically accessible, and safe to operate on. It's a protrusion of the fluid compartment, consisting of a thick-walled, flattened sac connected to the inner ear with a duct. Not surprisingly, the sac is called the endolymphatic sac, and the duct is called the endolymphatic duct.

In 1926, a French surgeon named Georges Portmann began performing operations on the endolymphatic sac to alleviate Ménière's disease.[12] The same way that glaucoma is a disease of excessively high pressures in the eye, he believed that Ménière's was a disease of excessively high pressures in the ear. In order to cure the disease, the pressures had to be lowered. By creating an opening into the sac, he believed that he was providing a pressure relief valve. Years later, it was also recognized that tumors growing out of the endolymphatic sac could cause Ménière's disease, further bolstering the idea that a dysfunctional sac was somehow involved. With my neurosurgery colleagues at UCSF, we even found that meningiomas—a benign and common brain tumor—caused Ménière's disease when they invaded the outer covering of the sac. Surgery on the endolymphatic sac is nearing its 100th birthday.

While endolymphatic sac surgery remains controversial to this day, there are certainly anecdotal tales of its success. In 1961, Alan Shepard Jr. became the first American to fly into outer space. However, he was benched from Apollo missions in 1963, as he developed attacks of dizziness and nausea, thought to be due to Ménière's disease. Seeking a cure, he learned of Dr. William House, a pioneering surgeon in Los Angeles, who is widely considered to be the father of modern ear surgery. In 1969, using his own surgical techniques, Dr. House made a small opening into Shepard's endolymphatic sac. The operation was hugely successful, and Shepard was allowed to return to active duty. In 1971, he commanded the Apollo 14 mission and made history as the fifth person to walk on the surface of the moon, and the first person to golf on the moon. Video clips of the lunar tee off are available online, where Shepard's three attempts at moon golf are recorded for posterity. The first two swings are not successful, and his crewmates and mission control can't help poking fun: "you got more dirt than ball that time"; "that looked like a slice to me, Al." Despite the one-handed swing forced by the stiff spacesuit, Shephard's 6 iron finally connects to the golf ball. He reports to mission control that the ball flew "miles and miles and miles" across the lunar landscape. He wrote a letter to Dr. House shortly before his death, thanking him: "I could not have done it without you."

As if the vertigo, tinnitus, ear pressure, and hearing loss wasn't enough, in its later stages, Ménière's disease also can cause drop attacks. Patients described being pushed to the ground by an invisible hand, collapsing in seconds, without losing consciousness. I've seen videos of drop attacks (like one caught by a security camera), and they are dramatic. People hit the floor with force, crumpling like a speeding car smashing into a wall. A theory was put forth by a British physician in 1936, in an article entitled "The Otolithic Catastrophe," leading many to call drop attacks "The Otolithic Crisis of Tumarkin" after the author, Alex Tumarkin. He notes that normally, the utricle and saccule (both of the otolith organs) help maintain muscle tone. Tumarkin points out several experiments, where manipulations of the utricle caused animals to switch between two opposing muscle states: maximum extension, where every muscle is as straight as possible, resulting in an upright posture, and maximum flexion, where every muscle is as bent as possible, resulting in the body curling up into the fetal position. He theorized that a sudden apoplexy of the utricle could cause a complete loss of extensor muscle tone, resulting in a drop attack. He wrote, "the patient will double up and collapse backwards like an empty suit of clothes."[13]

In the evening of December 3, 1888, Vincent van Gogh cut off his left ear. As reported in the local paper, he then took the severed appendage to his local brothel, asked for Rachel, and presented her with the ear, saying "Keep this object carefully."[14] Some ear specialists, including Masao Yasuda in 1979 and Irving Kaufman Arenberg in 1986, have maintained that Van Gogh suffered from Ménière's disease, arguing that his automutilation was a dramatic attempt to relieve symptoms of pressure and tinnitus. I'm skeptical. Van Gogh's behavior at the time was erratic and unmoored. There are reports that his friend, fellow impressionist Paul Gauguin, fled Van Gogh's company that night, after he was threatened with the very razor used in the disfigurement. Van Gogh spent much of the last two years of his life in hospitals for the insane. In fact, his transcendent masterpiece—*Starry Night*—was painted during a stint at the asylum of Saint-Paul de Mausole. I've searched his letters—as hundreds of correspondences to his brother Theo survive in an online database—and can't find evidence of Ménière's. He doesn't describe tinnitus, instead admitting to auditory and visual hallucinations, writing "I observe in others that, like me, they too have heard sounds and strange voices during their crises, that things also appeared to change before their eyes."[15] Van Gogh was suffering from psychosis, perhaps precipitated by epilepsy, absinthe, syphilis, or some combination thereof. This led to his suicide in 1890, when he shot himself in the chest with a revolver. While to outsiders it may seem logical that some enduring the torture of Ménière's would do anything to ease symptoms, acts of self-harm point to manifestations of psychiatric disease. Van Gogh's story is reminiscent of one patient of mine. Suffering from schizophrenia, and wishing to quiet the voices in her head, she plugged both ear canals with superglue. Fortunately, in that case I was able to extract the hardened globs, with no permanent damage done.

I don't know what the future will hold for Ménière's disease. I don't understand how we will cure a disease until we understand it. There are glimmers of hope on the horizon: new ways of imaging the inner ear to see hydrops, anatomic variations that seem to correlate with disease features, and new clinical trials. I keep coming back to the curious fact: half of those with Ménière's disease have a migraine history. What if, somehow, migraine causes Ménière's disease?

10

Senseless

Bilateral Vestibular Loss

THE BIOLOGIC CINEMA SYSTEM

Some of the earliest medical accounts of bilateral vestibular loss come from Walter Dandy. Dr. Dandy was a pioneering neurosurgeon at the Johns Hopkins Hospital in Baltimore, active during the first half of the twentieth century. Among his many achievements, in the late 1920s Dandy developed an operation for Ménière's disease. His approach involved surgical destruction of the eighth cranial nerve, otherwise known as the audiovestibular nerve. At the time (and sadly still today), the cause of Ménière's was not known, and medical treatments were frequently not effective. Therefore, the operation involved cutting the vestibular nerves, which was either done on one side, or on both sides. Cutting the vestibular nerve can be thought of as permanently unplugging the inner ear. The hair cells may continue to try to send signals, but without the nerve, the signals can't be relayed to the brain.

Dandy published his experience in 1941, in the *Journal of Surgery, Gynecology, and Obstetrics.*[1] He wrote that "if one auditory nerve—or only its vestibular division—is divided there is no permanent loss of vestibular function of any kind." We now know that this is not factually correct, and that there is a permanent loss of a function after a vestibular nerve section. However, Dandy was likely describing things at the level of behavior, as he lacked the tools to properly study the function of the vestibular system. And

143

it is true that after a unilateral (just on one side, so right or left) vestibular nerve section (section is a medical term for division), most people compensate by using information from their unaffected side. So, even if they have a deficit, it may be subclinical, meaning that it does not affect their ability to function at all.

However, sometimes Dandy would perform his operation on both vestibular nerves. He wrote, "Division of both vestibular nerves is attended by one rather surprising after-effect, i.e. jumbling of objects (visual) when the patient is in motion; as soon as the patient is at rest the objects are again perfectly clear. The other disturbance is uncertainty when the patient is walking in the dark. . . . These two disturbances indicate the very intimate association between the vestibular and the visual apparatus in human beings." Despite his incomplete understanding of the function of the vestibular system, Dandy's description is quite accurate and is echoed by patients who suffer from bilateral vestibular loss.

It's worth understanding in detail why someone without a vestibular system would have those two issues. You will recall from chapter 4 that one of the primary jobs of the vestibular system is image stabilization. That means that every time we move our head, we need to adjust our eyes slightly so that they can stay focused on something. This system is not necessary at rest, because there is no head or body movement. But, without the vestibular system when we start moving, the world bounces around. Eyes can't keep up, and the world goes from clear to blurry. By analogy, try to take a picture with a phone camera in low light (with bright light, the capture time is quicker, and you may not see the effect). If you hold the phone very steady, you can get a reasonable picture. But, if the phone is moving quickly up and down, the picture will appear unfocused. Unfortunately, many patients with bilateral vestibular loss live inside that blurry world.

Without vestibular sensors, balancing becomes more difficult. In ideal conditions, the brain can supplant the lost information with visual and sensory input. In addition, deliberate and careful movement can help avoid falls. But, when the lights go out and the ground becomes soft or uneven, the brain simply does not get enough data to calculate position and heading. Without our failsafe internal compass—the vestibular system—we stumble.

In March of 1952, a personal account of bilateral vestibular loss was published in the *New England Journal of Medicine*, titled "Living Without a Balancing Mechanism."[2] At the time, the author's identity was only known as JC. JC, it turned out, was not only a patient, but also a physician. His narrative offers a firsthand perspective on the everyday travails of life without a functional vestibular system. JC's life begins to change about two and a half months after a series of streptomycin injections into his knee, for arthritis related to tuberculosis. The age of antibiotics only began in 1928, with the discovery of the original wonder drug: penicillin. Other antibiotics followed, including streptomycin in 1944. Like penicillin, it wasn't designed in a lab, instead it was found in nature. Both antibiotics, like many medical compounds, are produced by microorganisms as a protective mechanism against hungry bacteria. At the time, streptomycin was the only medicine that had any chance against tuberculosis.

JC recounts the first moment, when he realized something was wrong (and I ask the reader's indulgence for excessive quoting, but I find JC's words so compelling and insightful, that I dare not dull them with paraphrasing):

In preparation for shaving, I wrung out the facecloth in steaming hot water, spread it over my hand and then held it to my face. Thus blindfolded, I suddenly lost my balance and fell sprawling on the floor. During that first day symptoms increased rapidly. Every movement in bed now caused vertigo and nausea, even when I kept my eyes open. If I shut my eyes the symptoms were intensified. At first, I found that by lying on my back and steadying myself by gripping the bars at the head of the bed I could be reasonably comfortable. Later, even in this position the pulse beat in my head became a perceptible motion, disturbing my equilibrium.

Most of us have experimented with motion pictures at home. This experience can be used to illustrate the sensations of the patient with damage to the vestibular apparatus. Imagine the results of a sequence taken by pointing the camera straight ahead, holding it against the chest and walking at a normal pace down a city street. In a sequence thus taken and viewed on the screen, the street seems to careen crazily in all directions, faces of approaching persons become blurred and unrecognizable and the viewer may even experience a feeling of dizziness or nausea. Our vestibular apparatus normally acts like a tripod and the smoothly moving carriage on which the professional's motion-picture camera is mounted. Without these steadying influences, the motion picture is joggled and blurred. Similarly, when the vestibular influence is removed from the biologic cinema system, the projection on the visual cortex becomes unsteady . . .

I learned not to do certain things. One of these was not to look at a newspaper or letter in my hand while walking . . .

Indeed, I soon noticed that unconsciously I had begun to walk the same type of course that is steered by a ship's gyroscopic compass, veering first slightly to the left and then overcompensating and veering equally to the right. . . .

During a walk I found too much motion in my visual picture of the surroundings to permit recognition of fine detail. I learned that I must stand still in order to read the lettering on a sign. These early excursions taught me a habit foreign to one of my New England background—that of greeting anyone who happened to pass in the opposite direction. Since I was unable to distinguish the familiar from the unfamiliar faces when walking, the obvious solution was to pretend to recognize everyone.

Learning to get about at night or the darkness has been the most difficult part of the convalescence. Even after considerable practice in daytime walking, I still find myself almost helpless in the dark—so helpless, in fact, that at night I have sometimes had to resort to achieving my destination on hands and knees.[3]

Fortunately, JC did recover with time and was able to resume his favorite activities of research, teaching, and playing tennis. Unfortunately, many of those who suffer from vestibular loss never recover to that degree and are left with permanent disability. Currently, the main treatment for bilateral vestibular loss is physical therapy. However, as we'll see in chapters 12 and 13, the future appears bright, as scientists race to find ways to replace vestibular function, either through regenerative techniques, or with implanted prosthetic devices. So, pardon the delay, but we'll get to the exciting, new treatments at the end of the book. I wanted to end the book on a high note.

Amid all the bad, JC did notice one curious benefit of his condition: "On the water I am spared the sensation of seasickness and hence am a useful hand in the hot galley when the seas are choppy."

LOST AT SEA

Seasickness was rampant in the ancient world, and even Hippocrates, the father of medicine, commented that "sailing on the sea proves that motion disorders the body." Hippocrates is more famous for his eponymous oath, variants of which are recited by newly minted physicians each year. Of historical interest, the original formulation of the oath, as viewed through modern eyes, contains good (I will use treatment to help the sick), controversial (I will not give a woman a pessary to cause an abortion), and anachronistic (I will abstain . . . from abusing the bodies of man or woman, bond or free) elements. The strong connection between boat travel and dizziness is encapsulated in the word *nausea*—from the Greek word *naus*—of the sea.[4]

JC's observation that motion sickness is reduced or eliminated with vestibular damage has been corroborated by others. In fact, William James, the nineteenth-century Harvard professor and "Father of American Psychology," first noted that deaf-mutes did not appear to suffer from nausea while on a sea voyage. He later followed up with a larger study, employing "violent rotations," and confirmed that he was unable to cause seasickness in a large percentage of deaf subjects (presumably those who didn't get sick had no vestibular function, as many diseases destroy both hearing and balance). Furthermore, it holds true whether the cause of motion sickness involves self-movement, or just movement of the visual surround. In one study, Bob Cheung and colleagues found that if they placed subjects within a moving sphere, so that they were still but their entire visual field was rotating, most would get dizzy. However, subjects with bilateral vestibular loss experienced no such dizziness.[5] Why is that?

There are several theories that try to explain motion sickness. According to the sensory mismatch theory, motion sickness ensues when the brain tries to sort conflicting sensory information. In the example above, the visual system says, "we're moving," and the vestibular system says "no, we ain't." On a boat, the opposite can occur, especially when all you can see are parts of the boat moving with you. The vestibular system says, "definitely moving," and the visual system, analyzing the fact that your view relative to the boat hasn't changed, says, "no we ain't" (the entire vessel might be bobbing up and down, but you are too, so if all you see is boat, the boat looks still. Looking out beyond the boat, at the horizon, gives you the visual sense that you are moving also, which may explain why that is helpful). Without a vestibular system, there is no conflict, no mismatch, and therefore motion sickness does not occur.

Researchers have wondered—even if it's true that discordant information is the cause—why motion sickness involves feeling queasy and vomiting? Why doesn't it instead cause a headache, or a seizure? Similar to coughing, vomiting is considered a protective reflex, to expel partially digested harmful substances, prior to full absorption. If so, motion sickness may be the unintended result of a poison detection system. In the pre–Trader Joe's world of our genetic forebears, identification of safe food was an existential concern. But, in the world of biology, everyone competes to win, and many plants and animals evolved venoms, toxins, and banes to avoid ingestion. Since many of those compounds produce sensory hallucinations (think back to your college days, esteemed reader!), it follows that unreliable signals would be interpreted as a potential threat. Real movements—running, riding a horse, cliff diving—generate trustworthy streams of data, where visual, sensory, and vestibular information are concordant. Discordant input is blamed on bad berries, which must be removed posthaste. In support of this theory, Kenneth Money and Cheung found that if the vestibular system was removed from dogs, they would not vomit as often when fed various poisons.[6]

For many, motion sickness is an inconvenience. But for pilots and astronauts, it can be lethal. If pilots become disoriented, they could crash their aircraft. And if an astronaut vomits in a spacesuit, they could asphyxiate. Forgive the mental image, but bear in mind that without gravity, the vomitus floats, and it could block the breathing passages. As related in his book *An Astronaut's Guide to Life on Earth*, Chris Hadfield, former commander of the International Space Station, was temporarily blinded during a spacewalk. Antifog solution—normally used to keep the inside of the visor clear—floated into his eye. Tears welled up, but without gravity they pooled around his eyes, making his vision worse. In the space station, this problem is easily solved by wiping the eyes. However, in a spacesuit, that isn't possible. His hands, unable to reach his face, were useless. Thankfully, the tear blob eventually disengaged, before backup plans (like venting air from the suit into space) had to be enacted.[7]

Understandably, the U.S. military has poured resources into studying motion sickness and disorientation. I once toured the Naval Medical Research Unit, in Dayton, Ohio. The state-of-the-art facility is the spiritual successor of the aeromedical research facility in Pensacola, Florida, which had a distinguished seventy-year history. For much of its history, it was led by Dr. Ashton Graybiel, who helped design spacesuits, pioneered a transdermal patch for scopolamine (a motion sickness medication), and studied the effects of extreme g-forces and weightlessness on humans.

The facility in Dayton was impressive, with a seemingly endless array of custom-built equipment: flight simulators, contraptions for safely dropping, spinning, and launching harnessed research volunteers, and complex arrays to capture 3D kinematic data. However, it was clear that the prized possession was a nineteen-million-dollar "disorientation research device," affectionately dubbed "The Kraken." Housed in a cavernous cylindrical room, it consists of a futuristic pod, suspended within arrays of concentrically larger frames, each able to spin, flip, or translate the pod about a different axis of movement. Test pilots are secured inside the pod, which is designed like a virtual cockpit. From a press release: "At 245,000 pounds, 4,500 horsepower, simultaneous motion on six axes, sustained planetary motion of 3 g, and horizontal travel to 16½ feet, the Kraken will allow researchers to create the most realistic motion simulations never before imagined."[8] Sadly, access inside the simulator was restricted to active military personnel. I tried informing the courteous staff that as I child, I once rode "The Great American Scream Machine," a now-defunct roller coaster, six times in a row, but this revelation did not earn me access.

Another problem sometimes arises from the unnatural environment of the sea. After being at sea, most develop sea legs, as they adapt to a world that rolls back and forth. An internal metronome marks the timing of each wave, so that each movement can account for an environment in motion. On return to dry land, the adaptation persists, as sailors stumble to and fro, anticipating waves that no longer exist. Thankfully, that period of adjustment is short, especially with experienced seafarers. However, with a rare vestibular condition, sufferers struggle to regain their land legs. They constantly feel a back-and-forth sway, as though they are rocked by invisible waves. The only reprieve is when they are in passive motion again, like riding in a bus. This disease is called Mal de Débarquement Syndrome (aka MdDS), from French for the Disease of Disembarking. At this point, there's no known cure for MdDS but treatments with vestibular physical therapy, magnetic stimulation of the brain, and migraine medications can help. In addition, there's some evidence that a special physical therapy treatment, developed at Mount Sinai Hospital in New York City, can help some sufferers. It involves having your head rolled side to side while inside a small cylinder whose walls appear to be rotating. Curiously, a large percentage of those with MdDS have a history of migraine headaches, and many get better with migraine treatments, raising questions about the relationship between the two disorders.

There are many causes of bilateral vestibular loss. They include genetic errors, toxic antibiotics, infections like meningitis, and degenerative neurologic diseases. However, half of all cases are idiopathic. In medical school, I was taught that the definition of "idiopathic" is a disease that makes doctors feel like idiots, because we have no idea what is going on. In a study of 154 patients with bilateral loss, researchers Florence Lucieer and colleagues found another migraine connection. In the half of subjects with no clear cause for their vestibulopathy, half had a migraine history.[9] Since the rate of migraine in the population is about 10 percent, this is unlikely to be a coincidence. Time and time again, migraine damages the inner ear in ways that we don't currently understand.

Bilateral vestibular loss isn't a common disease. However, for those affected, it can be devastating. Simple things, like walking to the bathroom at night, can become treacherous, the quiet hallway now a deadly obstacle course. Bilateral loss also paints a clear picture of the vestibular system. Only in its absence can we truly see how essential it really is.

11

A Hole in the Head

Superior Canal Dehiscence Syndrome

A MINOR DISCOVERY

Medical discoveries are inspiring. It's hard not to be wowed by the inventive minds who walked in the same world as others and saw something that no one else did. In 1847, Ignaz Semmelweis discovered that the simple act of washing hands with an antiseptic prior to delivering a child dramatically reduced mortality (Sadly, his theory was not accepted, and he eventually suffered a nervous breakdown, was committed to an insane asylum, got beaten by the guards, and died from his infected wounds). Today, of course, handwashing is a pillar of modern medicine. Alexander Fleming turned the tide of a silent war between us and hungry microorganisms when he discovered penicillin in 1928. Prior to then, infectious disease was the leading cause of death, afterward, it became cancer and heart disease.

New discoveries are also humbling, considering that they can uproot long-held beliefs and dogmas. We want to believe that we have all the answers, but the truth is that we don't. Sometimes we know what we don't know. For example, we don't currently understand Ménière's disease (at least at the time of writing in 2024). Scarier is the fact that we don't always know what we don't know. Which facts, earnestly taught to us, held dearly in our memories, are actually false? Any health care professional wonders what "truths" that they believe today will be obsolete tomorrow. Medicine, like science in general, is a self-correcting machine, and over time there is a general trend toward an expanding sphere of knowledge. But to progress forward, we must be willing to discard old and disproven ideas.

The state of the art in my field—neurotology—is quite different than it was one hundred years ago. Today, we implant bionic ears to restore hearing to the deaf, employ precise imaging scans to see inside the intact skull, and use microsurgical instruments and lasers to rebuild the delicate hearing mechanism. None of that was possible in the 1920s. During my fellowship training at Johns Hopkins, I spent hours looking through temporal bone specimens from the early twentieth century, along with case histories. Thankfully, many of their ailments—like rickets and tuberculosis—are now almost unheard of in the developed world. Other would-be causes of death—like mastoid suppuration—are quite infrequent because of antibiotics and modern surgical techniques. But we are not at the pinnacle of the mountain of knowledge, and I would give a kidney to see what ear medicine will look like in another one hundred years.

In 1998 (not *that* long ago) Dr. Lloyd Minor and colleagues published a groundbreaking manuscript, describing an unknown disease.[1] Prior to then, symptoms were misdiagnosed, disregarded, ignored, or, at best, categorized as unexplained. At the time, Minor was already a notable vestibular researcher and had published research on the neural mechanisms of the vestibulo-ocular reflex, based on experiments in squirrel monkeys. By 1995, Minor had noticed a pattern with a small group of dizzy patients. By applying puffs of pressure, or with sound, he was able to cause nystagmus (a rhythmic twitch of the eyes due to vestibular activation—see chapter 4 for a review). Not only that, but the pattern of nystagmus—the direction and plane of the eye movements—pointed to a clear culprit: the superior semicircular canal. You will recall that based on the physics, anatomy, and neural circuitry, specific activations or inhibitions of the different parts of the vestibular system determine the reactionary eye movement. Minor, a vestibular detective, armed with knowledge of the principles governing nystagmus, was able to deduce their cause.

He found that similar to Benign Paroxysmal Positional Vertigo (BPPV), the vestibular system was being fooled. This time, it was not loose crystals causing the hair cells to bend, sending their spurious signal to the brain, resulting in the illusion of movement—aka vertigo. Instead, now it was sound and pressure. It was long known that pressure changes in the ear—produced in the clinic by gently squeezing a rubber bulb attached to a pneumatic seal at the ear canal entrance—could cause vertigo and nystagmus in some patients. Camille Hennebert, a Belgian otolaryngologist, first described this in 1911, and today it's known as "Hennebert's Sign." Hennebert's sign occurred in patients with syphilis, which was quite common prior to antibiotics, affecting the likes of Lenin, Oscar Wilde, and Manet. Pietro Tullio, an Italian biologist, described his eponymous sign based on experiments in pigeons. By drilling little holes into their semicircular canals, and playing sounds, he was able to cause nystagmus.

Dr. Lloyd Minor had discovered a new disease. So, the first treatment for superior canal dehiscence was clearly in the late 1990s. But when was the first case? It's an intriguing question because superior canal dehiscence results from an anatomical error. Unlike some other medical maladies that affect soft tissue (a surgeon's term for everything squishable in the human body), this is a bone disease. Not only that, but the dehiscence occurs in the densest bone in the human body. A German research team may have stumbled on the answer while studying the skulls of Egyptian mummies.[2] They analyzed the University of Marburg collection, composed of ten heads from the regions of Abydos, Thebes, and Philae, ranging from 1,700 to 5,000 years old. They found one specimen with an incredibly thin bone (less than 125 microns) over one of the superior canals. Unfortunately, no details were available for that individual's life (other than sex), so we don't know his symptoms. However, based on the bone thickness, it is possible that an ancient Egyptian did have a form of superior canal dehiscence during life, prior to being mummified.

WINDOWS OPERATING ERROR

Patients with superior canal dehiscence—just like some with ear syphilis, or those with artificial holes in their inner ear—are subject to an unfortunate symptom: loud sounds and pressure changes can make them dizzy. My patients have reported a variety of triggering sounds: pots clanging, dogs barking, babies screaming, passing sirens. There are some who are sensitive to all sounds, even the sound of their own voice, so they whisper to avoid creating a perceptual earthquake. They have also reported a variety of everyday activities that cause ear pressure changes (and therefore dizziness): lifting weights, coughing, sneezing, flying, or trying to go to the bathroom while constipated. A common thread of those examples is that they cause a transient pressure spike in the brain cavity or middle ear space, which then spreads to the inner ear, mimicking a sound wave. But why would sound make someone dizzy?

To answer that question, it's useful to start with the opposite question: Why doesn't sound normally make us dizzy? The inner ear is a shared workspace for the balance and hearing systems. We previously saw why—related to the evolutionary history of hair cells being conscripted for various useful biologic purposes. But there are no dividing walls within the inner ear. The stapes, which relays sound vibrations from the middle ear to the inner ear, sits atop the vestibule, where the otolith organs are. Anatomically, the stapes is closer to the saccule (sensing gravity) than to the cochlea (sensing sounds). How does sound "know" that it's supposed to go to the cochlea, rather than to the semicircular canals?

The answer relates to the anatomy of the inner ear and principles of fluid dynamics. Simply stated, sound waves follow the path of least resistance. The inner ear is encased in the hardest bone in the body. And it's filled with fluid, which is relatively incompressible. When the stapes is pushed inward, that force needs an outlet. Otherwise, the fluid in the ear would push back so hard that functionally, the stapes wouldn't move, and sound waves would dissipate at the stapes footplate—the interface between the middle ear and inner ear. The stapes is shaped like a stirrup, and the footplate is at its bottom. The stapes footplate is held in place by a circumferential ligament, allowing it to bounce while still sealing off the inner ear. The bony gap where the stapes footplate sits is called the oval window. Therefore, the inner ear does have a second opening. It's a springy membrane, situated in a protected bony alcove, called the round window. Every time that the stapes footplate is pushed in, the round window bulges out. This phenomenon is called the round window reflex (it's not a true reflex because it doesn't involve a neural circuit, but it's called a reflex anyway. It's more of a mechanical certainty. And it's not the only ear reflex that isn't truly a reflex. When looking inside an ear canal at the eardrum, there is usually a triangular swath of reflected light toward the front of the eardrum, just due to its shape. That reflection is called the light reflex. For those keeping track, I did not include either of these pseudo-reflexes in the fastest reflex contest at the beginning of chapter 4). The round window reflex is quite useful to surgeons performing the delicate task of rebuilding broken hearing bones. When placing prosthetic replacement bones, one can check their alignment by gently compressing the prosthesis and looking for the round window reflex. If it's there, then the prosthesis is in a good position to transmit sound waves to the inner ear.

Sound waves travel to the cochlea, rather than other parts of the inner ear, because the cochlea lies between the stapes (the oval window) and the round window. Again, the remarkable inner ear solves problems with simple and elegant design. Even though the vestibular organs are in close proximity to the stapes/oval window, they are not in the path of least resistance, and therefore they don't get activated by sound pressure waves. In ear parlance, the oval window is also called the first window, and the round window is called the second window. So, the inner ear is a house of bone, with two soft spots that can balloon in and out, which are the windows. When Dr. Minor first described superior canal dehiscence syndrome, he blamed the symptoms on a "third mobile window." Now that we've covered how the system normally functions with two windows, let's see what happens when you open a third.

Just like opening actual windows on a windy day changes the flow of air through a room, new inner ear windows change the flow of currents through the labyrinth. Opening a new window opens a new stream for fluid waves to propagate along. Normally, there is just one pathway, from the oval window (stapes) to the round window. But if another window is open—say at the apex of the superior canal—then there will be a second pathway for sound energy to traverse the ear. A window, here, is any gap of the bony shell of the inner ear. With a superior canal dehiscence, bone erosion at the top of the canal results in an unroofed segment, a dehiscence. In that area, the canal is covered by the next layer up—which is dura—a tough, protective sheath that protects the fragile brain contained within. So, with a dehiscence, some of the energy will be diverted from the usual route to travel from the oval window to the third window.

This extra pathway for inner ear fluid ripples creates two problems. First, since some energy is shunted away from the cochlea, a mild or moderate hearing loss occurs. Second, fluid oscillations from sound are now crossing vestibular organs, like the ampulla of the superior canal. As they do, the waves shake sensitive hair cells, which are hardwired to interpret that as a head movement. Spurious signals are sent to the brain, causing nystagmus and vertigo. A similar thing occurs with pressure changes. For instance, with a cough, pressure inside the head transiently rises, which causes the dura to plunge inward, sending a shockwave through inner ear fluids, from the dehiscence to both the oval and round window membranes. With the third window, the mechanics of the inner ear are altered, and now sound and pressure can cause dizziness.

SUPER HEARING

There's a cacophony inside our heads. Brains pulsate with each heartbeat, making a whooshing sound. Eyes squeak as the dart around. Neck muscles and joints creak and groan. However, these sounds are not normally audible. Golden silence is an illusion, made possible by the fact that we simply don't hear the throbs, squeals, and rasps inside our heads. Even throat sounds—like voice, chewing, and breathing—all reverberate, but their volume is dampened by the closed Eustachian tube. With superior canal dehiscence, all those inaudible sounds are heard: annoying, distracting, and ever-present. Like many things, we don't appreciate the sound of silence until it's gone.

The exact mechanism behind the phenomenon of hearing internal sounds is not fully understood. These sounds don't use the normal hearing channels. Instead, they are conveyed directly through the bone of the skull, in a process known as bone conduction (normal hearing, conversely, is called air conduction, after the medium through which sound waves travel). Bone conduction bypasses the usual hearing machinery—the ear canal, the eardrum, and the ossicles—to directly vibrate the cochlea. The principles of bone conduction have been long known; Beethoven took advantage of it by affixing a rod to his piano and biting down on the free end to compensate for his failing hearing. (Author's note: of historical interest, it's not known why Beethoven lost his hearing, theories include stapes fixation from otosclerosis, intestinal autoimmune disease, and nerve damage due to drinking cheap wine tainted with lead). Researchers Saumil Merchant and John Rosowski put forward a hypothesis for the enhanced bone conduction seen in superior canal dehiscence.[3] When the entire cochlea vibrates, internal fluids quiver by displacing the oval and round windows, similar to how a shake weight works. The magnitude of the fluid wave in the cochlea determines how loudly (and at what loudness) a sound is heard. Normally, the oval window has much more resistance to movement because of the stapes sitting atop it than the round window, which is only a membrane and therefore easier to move. The difference in resistance to movement is a good thing, because if the oval and round window had the same resistance, then both would move in tandem with vibration, canceling each other out. Merchant and Rosowski argue that a superior canal dehiscence, directing energy away from the oval window, increases the difference in resistance between the oval and round window, making bone-conducted sounds louder, and therefore audible.

One of my mentors, Dr. John Carey, refers to superior canal dehiscence as "The Great Masquerader." It's a borrowed term, as the original masquerader was syphilis, well-known for its protean manifestations. Within the realm of ear disease, superior canal dehiscence certainly deserves the moniker, as it can present with just one of its various symptoms, and frequently there are many other diseases that can cause each symptom. Hearing your heartbeat, for example, can be caused by aberrant arteries and veins around the ear, increased pressure within the head, and certain tumors with a robust blood supply. The exception to that rule is hearing eyes move, which seems to be specific to superior canal dehiscence. In fact, many frustrated patients have successfully diagnosed themselves with a Google query for that curious symptom.

PUTTING IN A PLUG FOR TREATMENT

Not all patients with superior canal dehiscence need or want treatment. It's a disease that causes symptoms but otherwise is not directly harmful to the body. Therefore, in my practice, it's the patient who makes the decision to proceed with treatment. Treatment is relatively unsophisticated: one of the pipes has sprung a leak, so it needs to be patched.

Part of surgery school (also called residency or fellowship) is learning the safe corridors that can be created to different parts of the body, while minimizing collateral damage. For the superior canal, there are two approaches. The first was described by Dr. Minor in the original paper, which involves creating a rectangular window into the brain cavity just above the ear, and skimming the top surface of the temporal bone until the dehiscence is encountered. This is called the "middle fossa" approach, named after the vault in the cranium that is traversed. There are some nice features of this technique, including a direct view of the dehiscence which can confirm diagnosis, avoidance of bone drilling near the delicate inner ear membranes, and precise placement of repair material. The main downside is that risks of opening the skull (craniotomy) involve damage to the brain. Fortunately, that risk is quite rare, although like many surgical risks, it's not quite zero.

The second passageway to the dehiscence is through the mastoid, which is the bone protuberance behind the ear. Even though it's in the skull, mastoid bone is under the brain cavity. Mastoid surgery is quite familiar to ear surgeons, who spend countless hours navigating its network of connected air cells. During mastoid surgery, partitions between these small chambers are removed, resulting in a larger, merged air cell, which does not appear to have ill effects on the body. A skilled surgeon can safely chart a course through the air cells to the superior canal, avoiding myriad dangers along the way, like the nerve that controls facial expression. This is called the transmastoid approach. Its advantage is that there is almost no risk to the brain, and ear surgeons are very accustomed to mastoid surgery. The main disadvantage is that the location of the dehiscence is not precisely known. Instead, the surgeon must drill through the bone of the canal in order to locate the hole. The usual approach is to create two tunnels based on estimates of the dehiscence location—one in front, and one behind, in order to bookend the diseased area. Since both approaches—middle fossa and transmastoid—involve manipulating the inner ear, they both carry the same risk of inner ear damage: vertigo, dizziness, tinnitus, and loss of hearing.

Once the dehiscent canal is reached, the absent bone can be replaced with other materials, which is called resurfacing, or the inside of the canal can be occluded, which is called plugging. Intuitively, resurfacing is more appealing, as the missing part is replaced. In addition, when the canal is plugged, the natural function of the canal—sensing quick downward head turns—is lost. However, early experience at Johns Hopkins showed that resurfacing alone led to recurrences of symptoms. Therefore, many surgeons accept the loss of function of the superior canal on one side (1/6 of the canals) as a necessary sacrifice and perform both a plugging and a resurfacing. The calculations become more complex with bilateral disease (affecting both superior canals), as the combined loss is harder to compensate for.

Surgery is considered to be generally effective. In a study of long-term outcomes, I (together with colleagues) asked ninety-three patients who had surgery years ago about their experience. Of those, 95 percent said that if they had a friend with the exact symptoms that they had, they would recommend surgery as a treatment. The other 5 percent would not. Ninety-five percent is considered a very good approval rating for a surgery.

Sometimes, a career in surgery seems like a Faustian bargain. In exchange for a decade of dedicated study, sleepless nights, meager pay (medical school costs a fortune, and I personally got paid less than minimum wage during residency), and agreeing to an arbitrary system called the match where you relinquish geographic control over your life (I'll have you know, dear reader, that a very pretty girl dumped me after learning of a five-year conscription to St. Louis), you are granted the privilege of caring for your fellow humans. It's not always glorious, but there is a solemn pride that you permit yourself to enjoy when it seems like you've really helped someone. About a year after a superior canal operation, I received an email update from one of my patients, accompanied by two photos. In the photos, my patients wore a Minnie Mouse bow, her husband sported a Donald Duck shirt, and the kids were princesses. In the background, Disney's Magic Kingdom was visible. She wrote, "I just wanted to thank you again for giving me my life back." I was so touched. The bargain was clearly worth it.

PART 4

THE FUTURE

12

Restoring a Sense of Balance
The Vestibular Implant

THE BIONIC MAN

I first became interested in otolaryngology during a lecture on cochlear implants. The cochlea is an analog to digital converter, translating incoming sound waves into intricate patterns of electricity pulsed out along the cochlear nerve. When the cochlea fails, the rich world of sounds around us—the pitter-patter of a puppy's paws, the vibrating, vivid voice of a violin, the sonorous symphony of splashing surf—all fall of deaf ears. A cochlear implant replaces the biologic function of the cochlea with an electronic ear. Sounds are still "heard," but with a microphone instead of an eardrum. Those sounds still get sorted by pitch and loudness, but with algorithms, instead of resonant properties of different regions of cochlea. And those sounds are still relayed to the cochlear nerve, but with direct electric current, rather than with an elaborate chemical chain reaction.

In my mind, the future had arrived. I grew up watching *Star Wars*, *Robo-Cop*, *Blade Runner*, and *The Terminator*. The bionic ear was clearly the first step toward the bionic person. Instead of using medicines to repair damaged body parts, we could use devices to replace them. From a practical standpoint, the human body is a machine, designed to pump blood, rich with oxygen and sugar, to an ever-hungry brain. Individual parts can be replaced as needed, as long as the brain—the seat of consciousness, memory, and personality—is preserved. If Krang, foe of the Teenage Mutant Ninja Turtles, could do it, why couldn't we? I wanted in.

The cochlear implant is arguably the most successful of all human implants. More than half a million have been performed worldwide. The average recipient is transformed, from not being able to hear the angry chirping of street traffic to being able to hold a conversation in a quiet room. There's certainly room for improvement in sound quality—perception of music in particular is poor—but other implants don't seem to come close in sophistication or functionality.

Artificial arms and legs improve every year, and the days of the wooden leg and the hook arm are long gone. Modern prosthetics use high-end materials and physics to enhance function. Some can even be directly controlled by measuring brain, nerve, or muscle activity, and translating that into complex hand or leg movements. Currently, control appears crude, but with enhanced detection and improved processing, the future appears bright.

The retinal implant is perhaps the closest analogue to the cochlear implant. The retina is the eye's sensor, a curved array of light and color sensing cells at the back of the eye. Without a retina in good working order, we cannot see. The same way a cochlear implant transcribes sound waves into a meaningful pattern of electric impulses in the cochlear nerve, the retinal implant transcribes light waves in a meaningful pattern of electric impulses in the optic nerve. The optic nerve then transmits those signals to the visual cortex, where information is processed for recognition, navigation, balance, and behavior. Current devices can restore light perception, and a rudimentary sense of vision. While it may seem like a modest victory, for a user who can now tell the difference between the sidewalk and the street, it can be life-changing.

There are some other human-cyborg parts under development. Olfactory implants would be able to restore a sense of smell. Future versions of a voice prosthesis could read brain regions responsible for making sound and generate spoken language. And perhaps in the future, prosthetic hearts, lungs, kidneys, livers, and pancreases will be able to outperform transplanted counterparts. Currently, all those organs function better when transplanted from another person. However, organ transplantation requires strong medications to suppress the immune system, so that the foreign organ isn't recognized and destroyed. An artificial heart or liver, constructed from biocompatible materials, would not be so constrained.

One interesting question to ponder: What happens when implants not only replace human abilities, but surpass them? In 2012, Oscar Pistorius became the first double amputee to compete in the Olympic games (which is inspiring, unfortunately he later was convicted of murdering his girlfriend). At some point in the future, a retinal implant recipient might be able to see in infrared, like a pit viper. A cochlear implant recipient might be able to hear ultrasonic frequencies, like a bat. Or radio waves, like my car. Biology is remarkable, but it does have constraints: materials (mostly organic), composition (mostly proteins), and fitness (from an evolutionary standpoint, every adaptation must pay for itself by increasing the odds of successful reproduction). Biotechnology, on the other hand, has fewer constraints. Implanted components must fit well in the body, resist infection, be composed of compatible materials, and ideally not require batteries.

The success of the cochlear implant naturally led to the question of the vestibular implant. If a device could restore a sense of hearing, why not a sense of balance? Instead of encoding sound waves, the device would have to encode vestibular information, like the position and acceleration of the head. From a design standpoint, however, there are a couple of key differences between the cochlear implant and vestibular implant to consider. First, the cochlea only has one part—the cochlea. The vestibular system, on the other hand, has five different parts: the three semicircular canals, the utricle, and the saccule. As we'll see, different research teams around the world have adopted different strategies for vestibular stimulation, but none are currently attempting to restore function to all five parts.

Another difference arises when you consider the anatomic spacing necessary to encode distinct signals. Current technology literally relies on electric current (couldn't help myself there). However, at the scale of the inner ear, with the level of current necessary to spark the nerve, the current spreads out over several millimeters. This explains why, among cochlear implant manufacturers, the number of active electrodes in the ear differ from twelve to twenty-two, with no appreciable difference in performance. Having an array with 1,000 electrodes, even if it were possible to manufacture, may not increase performance. The electric signal is inherently fuzzy. In fact, it's been calculated that with current technology, no additional performance gains are realized beyond eight electrodes. In the cochlea, the advantage of a sharper signal would be clearer, more natural sound, with enhanced ability to tell the difference between similar pitches (like adjacent keys on a piano).

If the cochlea is in the middle of a difficulty spectrum for electric stimulation, the vestibular system lies on both extremes. The semicircular canals are easy—all the nerves fire together, based on the speed of acceleration. You will recall that all the hair cells share orientation, so any head movement that activates a hair cell will also activate its neighbors. Like an eighteenth-century British infantry unit, the hair cells and the soldiers are all facing the same direction, and they fire together. Because each semicircular canal encodes information about a single vector of movement, the replication process is relatively straightforward. The right horizontal canal nerve only fires when the head turns to the right. The left superior canal nerve only fires when the head is turned downward, to the left. To replace one set of semicircular canals, you'd need three rotational accelerometers, each placed in the exact same plane as a semicircular canal. Each accelerometer would then be connected to an electrode inside each of the canals. They would activate depending on the speed of acceleration, so slower movements get less current, and quicker movements get more. Now, of course, real-world head movements don't only occur along the axis of each semicircular canal. It would be rare for me to move my head so perfectly horizontally that I have 100 percent horizontal canal activation, and zero percent superior and posterior canal activation. But that's ok. The vestibular centers in the brain are used to getting a complex signal consisting of differential activation of each canal and translating that into a precise reading on the vector and speed of a head turn. For example, turning your head straight up to look at the ceiling at ten degrees/second might result in 50 percent activation of both posterior canals, 50 percent inhibition of both superior canals, and zero percent activation of both horizontal canals. So, anatomic separation of different canals, coupled with their relatively simple patterns of activation, theoretically implies that electrodes should be able to duplicate the precise neural signals of a normal vestibular system. (I am simplifying a bit here. Individual nerve fibers for each semicircular canal do have different properties, for example the speed of head movement at which they respond most vigorously. However, the general idea holds true).

However, the otolith organs are difficult. The saccule and utricle are tiny, just a few millimeters across. In order to sense any direction of movement or tilt, hair cells are all oriented toward a central dividing line, called the striola. Instead of all facing the same direction—like an audience watching a play, where everyone is facing forward to the stage—the hair cells all face in different directions—like in a football game, where everyone is facing the field, but the angle of looking depends on their position in the stadium. Effectively, that means that any electrode placed on the utricle or saccule would activate the whole organ due to the spread of current. Submillimeter precision, necessary to activate one part of an otolith, but not another part, is simply not possible with the technology of today. For the saccule, cells that fire when you move straight upward, and the cells that fire when you move straight downward, would both be activated. Therefore, with electric stimulation, any signals sent through the utricle or saccule would be inherently ambiguous and would not reflect activation patterns of any real-world stimulus.

There's a third critical difference between the cochlea and the vestibular system. When electrically stimulating the ear, especially with the help of an external device that could fall off or lose charge, we must consider what would happen if this stimulation stopped. For the cochlea, stopped stimulation means that you would no longer hear anything. Frustrating, but not necessarily dangerous. Keep in mind that in the United States, hearing is not required for a basic driver's license. The vestibular system is different. The normal vestibular nerve has a constant rate of firing, which then increases or decreases in response to head movement. When the nerve suddenly stops working, and one side goes quiet, the brain doesn't interpret that as no movement. Instead, it looks at the difference in firing rates between the two sides of your head, and computes that you must be spinning around. Really quickly. Keep in mind that if you were actually spinning, one side would increase firing (e.g., from 100 spikes/sec to 150), the other side would decrease firing (e.g., from 100 spikes/sec to 50), and the difference would be the same as the difference (100 spikes/sec) between baseline firing (e.g., 100 spikes/sec) and zero (also 100 spikes/sec). In response, you get vertigo, with a subjective sense of being on the merry-go-round, and reflexive eye twitching to compensate (nystagmus). Therefore, sudden malfunction of a vestibular implant might result in disabling vertigo. We'll examine what some research teams have done to try to solve this issue.

FROM EAR TO EAR

In the early 1960s, Bernie Cohen and colleagues showed that electric stimulation of the ampullary nerves—the individual nerve for each semicircular canal—caused predictable eye movements in cats.[1] This was a direct extension of J. Richard Ewald's prior research, which showed that activation of each canal produced an eye movement in the same plane as the canal. Ewald has used a pneumatic hammer, which was able to activate each canal with a gentle squeeze. Cohen was using electricity. By harnessing the vestibulo-ocular reflex (VOR), the researchers were able to move the eyes in any direction they chose. It's no coincidence that we have six semicircular canals and six eye muscles in each eye socket. As part of a lean design to minimize processing time, each canal is roughly in the same plane as an eye muscle. And just like the semicircular canals, eye muscles are organized in pairs, to pull the eye in either direction (muscles only contract, get shorter, so there always needs to be paired muscle on the other side to pull back and reset position). For example, the right horizontal canal is in the same plane as the medial and lateral rectus muscles. So, when you turn your head to the right, activating the right horizontal canal, the VOR tells the right medial rectus and the left lateral rectus muscles to contract, which moves the eyes to the left. With the head moving right and eyes moving left, there is no overall change in the direction of gaze, which is the goal of the VOR.

Cohen and his team at Mount Sinai Hospital in New York City were able to decode the VOR. By applying electric current to various combinations of the semicircular canals, and recording eye muscle contraction and eye movement, they assembled a library of instructions of how to get the eyes to move in any direction. For a pure downward movement, activate both posterior canals. For a pure upward eye movement, activate both superior canals. To get both eyes to roll to the right, activate the superior and posterior canals on the left, but not the horizontal canal. As expected, if you activate any of coplanar canals together (three pairs: both horizontals, left superior and right posterior, and right superior with left posterior), you get no eye movement. They cancel each other out since each one is giving the exact opposite directions to the eyes through the VOR. This landmark paper laid the foundation for future research, showing that the nerve for each semicircular canal could be stimulated by electric current in predictable patterns, mimicking that natural activation that occurs with head movements. Cohen and colleagues had created a vestibular codex, able to translate stimulation parameters to precise eye movements by hijacking the VOR.

Progress on the vestibular implant continued, albeit slowly. In the year 2000, researchers Daniel Merfeld and Wangsong Gong reported that they had developed a prototype vestibular prosthesis, and that in animal experiments, it worked.[2] This early device could only sense movement in one axis. There was a baseline stimulation rate, which then increased or decreased

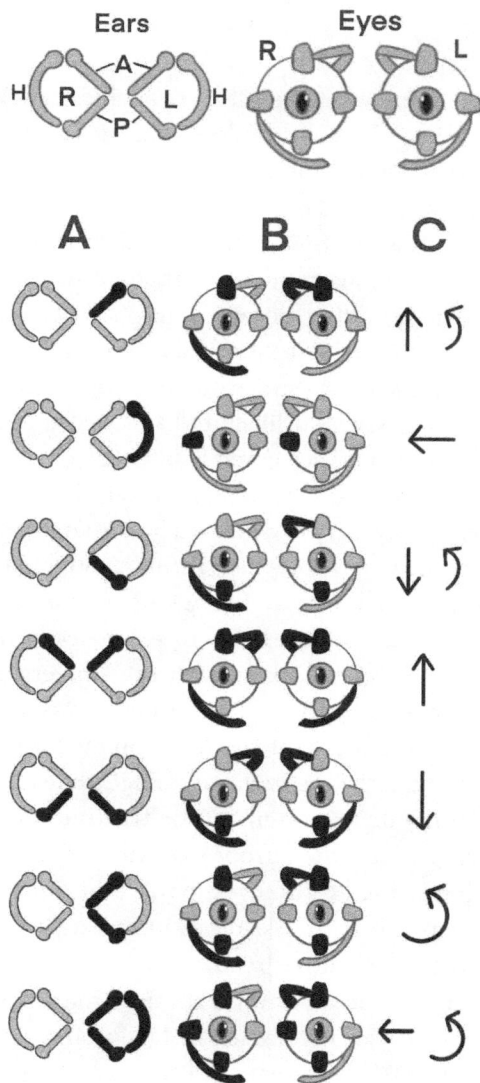

FIGURE 12.1 The vestibular codex. Different patterns of semicircular canal activation (*column A*) cause predictable patterns of eye muscle movement (*column B*). Column C shows the net movement of the eyes. The top row provides anatomical orientation. The six semicircular canals are shown in a simplified schema, with a top down view. *R* is right, *L* is left, *H* is the horizontal canal (lateral canal), A is the anterior canal (superior canal), and *P* is the posterior canal.

The eyes are shown in the front view, and the six eye muscles are not labeled. Electrical stimulation of each canal is represented in black shading, and the resultant eye muscle activation is also represented with black shading. The arrows show the net movement of the eyes for each pattern of activation. If two opposing eye muscles are activated, they will cancel each other out, and therefore are not represented in the arrow showing overall movement. For example, when both anterior canals are stimulated, the opposing actions of the superior and inferior oblique muscles cancel each other out because they rotate the eyes in opposite directions. Therefore, only the superior rectus muscles contribute to the upward eye movement.

as the internal gyroscope was rotated. As expected, increased stimulation of a single canal activated the vestibulo-ocular reflex and produced compensatory eye movements. Importantly, they showed that prolonged stimulation did not result in damage to the vestibular system. In another experiment, they also showed that after repeated cycles of turning the prosthesis on and off, the animals got quicker at adapting to having the device on. This was an exciting finding, because it implied that over time, humans could get used to the inevitable times when a vestibular prosthesis would fall off, run out of battery, or otherwise malfunction. With the first few ON/OFF transitions, subjects would likely be launched into a severe bout of vertigo, but after some time, they could habituate.

Over the next few decades, several teams around the world began doing the preparatory work necessary to achieve a human implant. At Johns Hopkins in Baltimore, Dr. Charley Della Santina assembled a large team of biomedical engineers, scientists, and surgeons to work on a multichannel vestibular prosthesis that would be able to sense head turns in any direction. A collaborative European team, with the Universities of Geneva in Switzerland and Maastricht in the Netherlands, worked on a modified cochlear implant, where a few of the cochlear implant electrodes were diverted for placement in the vestibular system. Unlike the Hopkins device, which is a standalone device intended for treatment of vestibular loss, regardless of hearing status, the European device was intended to simultaneously treat both hearing loss and vestibular loss. One advantage of that approach is that one of the main risks of vestibular surgery—hearing loss—is avoided, because the recipient is already deaf. A downside is that the pool of potential implantees is more limited because many have vestibular loss without hearing loss. A third team, led by Jay Rubinstein and James Phillips at the University of Washington in Seattle, took a different route. Rather than trying to use a vestibular prosthesis to restore vestibular sensation, they designed a device to try to treat Ménière's disease. They reasoned that directed electric current would be able to counteract aberrant signals during an acute attack, lessening or eliminating vertigo (those trials were eventually discontinued, subjects lost hearing and most did not have further attacks of Ménière's disease after surgery. It's possible that the surgery was causing enough damage to the vestibular system to cure Ménière's disease, similar to the way that gentamicin works).

The first human vestibular implant was performed in 2007, at the University of Geneva.[3] Prior to the implant, the surgical team had been experimenting with applying current to the semicircular canal nerves (aka ampullary nerves) under local anesthesia.[4] These patients were undergoing medically indicated ear surgeries, like cochlear implantation for deafness. During the surgery, a short detour was taken (with the patient's prior permission, of course), and a specialized electrode was used to stimulate one of the ampullary nerves for up to a minute. Instead of placing the electrode within the semicircular canal, the electrode was placed outside the canal, adjacent to the ampullary nerve, using a technique developed by Richard Gacek for the treatment of Benign Paroxysmal Positional Vertigo (BPPV, loose ear crystals). The research team was able to show that the posterior canal ampullary nerve was able to be stimulated, resulting in reflexive eye movements. They then repeated this approach with the other ampullary nerves, with similar results.

Based on the proof of concept, Jean-Philippe Guyot and colleagues proceeded to implant a patient with bilateral vestibular loss, for the first time, with a permanent electrode in the vestibular system. They partnered with the cochlear implant manufacturer Med-El, which is based in Innsbruck, Austria. Just to clarify, Med-El is different than Kal-El. Kal-El is Kryptonian for "Star Child," and was the name given to Superman as he was jettisoned from Krypton, fleeing the planet's destruction in an intergalactic escape pod. Med-El is short for "Medical Electronics." So, if you see General Zod attacking planet Earth, and you don't want to kneel, you'd best get assistance from Kal-El. But, if you wanted to develop a biomedical implant and get it past regulatory hurdles while still focusing on a viable business plan, then Med-El would be a better bet.

A typical Med-El cochlear implant has twelve active electrodes. One of those electrodes was separated from the standard electrode array and placed within a separate flexible arm coming off the implant. During surgery, the cochlear electrode array (with eleven electrodes) was placed into the cochlea for hearing restoration, and the extra lead was placed adjacent to the nerve for the posterior canal. Once the patient healed from surgery, the implant was turned on. The vestibular electrode was controlled by custom software in the laboratory, so it was only active during research sessions. The team found that when the electrode was turned on, reflexive eye movements, nystagmus, were seen. With continuous stimulation, the nystagmus slowly died down, over about half an hour. Furthermore, they found that by modulating the stimulation parameters, they could make the eyes wiggle back and forth. Unexpectedly, the eye movements were not vertical, as would be expected for posterior canal stimulation, but instead horizontal. However, the research team had broken through several barriers, and shown for the first time that you could safely implant an electrode into the human vestibular system.

The Geneva-Maastricht team continued to push the field forward, with a number of firsts. A multichannel implant was developed to stimulate all three canals. This was implanted in 2012. Further research showed that the implant could restore the vestibulo-ocular reflex. Experiments then showed that the implant was also helpful in improving vision during head movement, and that the vestibulo-ocular reflex could be restored even for the very quick head movements that occur during everyday life.

Meanwhile, at Johns Hopkins, Dr. Charley Della Santina and his team were diligently working, solving one problem after another on the road to vestibular implantation. I had the privilege of training under Dr. Della Santina for two years during my fellowship. He's one of the few true geniuses I've ever met. He's a visionary and recalled a brainstorming session in 1999 with Dr. Lloyd Minor (who we met in chapter 11) where they were discussing unmet needs in the field while waiting for a taxi in Miami. Lloyd asked Charley, "Why not a vestibular implant?" But great accomplishments need more than just inspiration. Luckily, Dr. Della Santina is driven by his singular goal. He had the fortitude and grit to push forward, day and night, for over twenty years. He's also a true scientist, meticulously checking every aspect of his creation with controlled experiments. He told me that he had two main motivations: he loves helping his patients, and he also loves solving the puzzles that land at your feet every day when you take on a gargantuan project like developing a new implant for human use.

I asked Dr. Della Santina what he found so interesting about the vestibular system. He told me, "It's a normally silent system that's working all the time to facilitate the things that we perceive, and yet, when it's working normally you don't even know it's there. And so, in a way, you could see an analogy to the Matrix, or to the ether, or some pervasive layer of truth that underlies the reality that you experience, and when there is a glitch in it, then you have vertigo, you perceive some weird movement in your vision, but otherwise it's a completely silent system that people don't even notice. Its absence is very noticeable. Its presence is something that in English we don't even have a word for."

After years and years of animal research, the first human implant was performed by Dr. Della Santina and his team in 2016 (actually just a few months after I graduated from my fellowship). This device was developed in partnership with Labyrinth Devices, LLC (which was founded by Charley) and Med-El, but unlike the European device, it was a dedicated vestibular implant. Instead of having just one electrode in each of the semicircular canals, three were placed in each canal, to allow the programmer to select the optimal electrode for stimulation. Due to the close anatomic proximity of the superior canal and the horizontal canal, those two electrode arrays were joined in a fixed shape that looks like a carving fork. Relatively speaking, the posterior canal is far away from the other two, so it got its own array. Similar to a cochlear implant, there was an internal device, which is surgically placed under the skin, and an external device, responsible for processing information and giving commands to the internal device. Both the internal and external devices have a magnet, ensuring coaxial alignment of the communication coils. Due to fears that a sudden disconnection of the vestibular implant could result in vertigo, it was bolstered with two additional magnets, for a total of three, guaranteeing strong connection at all times. At the time of writing (and I am certain this information will be out of date before this book is published) fifteen subjects had been implanted. All had severe loss of vestibular function in both ears, as confirmed with extensive laboratory testing. They each wore their devices 24/7, for prolonged periods of time (to date, each subject has continued to use their implant after surgery). Each only received a vestibular implant on one side, as expected for an initial trial where safety is a key concern.

Preliminary results were amazing.[5] The implant worked as intended, restoring vestibular function. Each time a subject moved their head, the three-axis gyroscope system would sense it, and activate the appropriate combination of semicircular canals, eliciting reflexive eye movements in the correct plane. Each subject really liked their implant, and they uniformly requested that they continue to wear their device beyond the required study period of two months. Many reported improved balance, improved quality of life, and a feeling that the life limitations imposed by their vestibular disease were lifted. Dr. Della Santina told me that he had expected the vestibular implant to improve vision during movement, but that he was surprised by the improvements measured in walking and balance.

As with any new technology, challenges and questions abound. Any inner ear surgery, including operations to replace the stapes bone, cochlear implants, repair of superior canal dehiscence, or the vestibular implant, all carry risk of causing permanent hearing loss. The inner ear is extremely delicate. While we think as surgeons that we are extraordinarily gentle with our hands, there is also a massive scale mismatch. The entire inner ear is about the size of a fingertip. A micromovement at human scale may be an earthquake to the ear. Fluid shifts can tear delicate membranes, causing permanent damage. Furthermore, even in the absence of direct trauma, inflammation can seep into the inner ear through surgical ports, which is the otologic equivalent of letting the bull loose in the china shop. Both stapes and superior canal surgery, which involve very small inner ear openings, carry a 1 to 2 percent risk of permanent deafness in that ear. Cochlear implantation, which is the closest analogue to vestibular implantation, involves placing a ~25 millimeter electrode array into the cochlea. Whereas the original recipients in the '90s had no measurable hearing, often today's implantees do have some natural hearing at implantation. They haven't lost 100 percent of their hearing, just enough so that they can't have a conversation, even with hearing aids. Under those circumstances, adult recipients lose any degree of residual natural hearing about 50 percent of the time. With the vestibular implant, some research subjects at both Johns Hopkins and the University of Washington lost some hearing. For the Geneva-Maastricht device, subjects already had hearing loss in the affected ear, so this was not an issue. Future work is needed to assess how to reduce, and hopefully eliminate, the risk of hearing loss with vestibular implantation.

When I spoke with Dr. Della Santina, he mentioned another problem that he called the informational bottleneck. He notes, "We really can only afford to have one electrode per canal, and we are trying with that one electrode to somehow convey the information that nearly 4,000 neurons convey. And those neurons are all different, and so with just one signal it electrically shocks all of the neurons the same way, that means we have an informational bottleneck. When we are talking to the brain through the vestibular implant, we are talking through a one-channel device per canal, instead of an 12,000 channel device, so it's like going from super high-speed Wi-Fi back all the way back to Morse code. With time, we'll technologically get better, we'll have more electrodes per canal, we'll figure out ways to change the current stimulus to selectively address different types of neurons, but for right now we are limited, as are developers of cochlear implants, pacemakers, and spinal cord stimulators, to a fairly small number of channels to control a very complicated system."

Currently, vestibular implants are only being used for patients with bilateral vestibular loss (both ears). For ideal function, do subjects need one implant, or two? Would patients with a vestibular loss in just one ear also be helped by a vestibular implant? Is it helpful for any other conditions? Should children with congenital vestibular loss be implanted, like how they are given cochlear implants for congenital hearing loss? Is it better to place the electrodes inside the semicircular canals, or outside the canals, directly next to the nerve? And what about the otolith organs? When will vestibular implants be commercially available?

After twenty years of sweat, dreams, equations, wins and losses, and the scientific process, the vestibular implant was born. In 2021, Della Santina and colleagues described the initial results in the most prestigious medical journal in the world: *The New England Journal of Medicine*.[6] Most of the tables and figures included are dense: tables of electrical stimulation parameters used during device activation, graphs of frequency-specific hearing thresholds measured in decibels. But there's an exception. Supplementary Figure 1 looks nothing like the rest. Instead, it's a photo collage. Each of the eight original recipients was asked to submit a photo of themselves, doing something that would have been impossible before the implant. It's hard to not be moved. The scenes are so simple, things taken for granted. Two women are seen gardening. One man, a former athlete, is running on a treadmill. Another climbs a ladder, the implant visible behind his ear. One trods through the snow. A smiling woman rides by on a blue bike, with a sign taped to the front: "Look, Dr. Charley, I Can Ride." And in the bottom right panel, beaming brightly, the eighth recipient is proudly dancing with his daughter on her wedding day.

Despite challenges, the future appears bright. Many with bilateral vestibular loss suffer, despite years of physical therapy. I believe that in the future, vestibular implantation will be a routine surgical procedure, just like a cochlear implant is today. There will be missteps, complications, and stumbles along the way. However, the basic principle, that the function of the semicircular canals of the vestibular system can be supplanted by a bionic system, has been proven.

13

Vestibular Rogaine

Hair Cell Regeneration

THE EARLY BIRD

Birds are amazing. The wandering albatross has a wingspan of up to eleven feet, can fly five hundred miles in a day, and is so efficient that it spends almost its entire life in the air, landing only to mate. The peregrine falcon can reach speeds of 200 miles per hour during a hunting dive-bomb known as a stoop, careening with such force that prey are killed instantly when punched with the falcon's clenched talons. Rüppell's vultures have been sighted gliding at 37,000 feet of elevation, otherwise known as the cruising elevation of a commercial jetliner. At that height, the curvature of Earth is visible. Birds talk, sing, build homes, get married, mourn, dance, flirt, steal, trick, act, imitate, and laugh. They are found in every environment on Earth, inhabiting land, sea, and air. But for the vestibular science community, there's something even more amazing about birds. They can regenerate their vestibular system.

And it's not just birds. Almost all nonmammals that have been tested can regenerate their inner ears to some degree. They can regrow dead or damaged hair cells. So, it's basically everyone but us. The obvious question is why? Isn't regeneration a good thing? If evolutionarily simpler species have that ability, why did we lose it? Is there a downside to regeneration? Can we harness the mechanisms used by other species to regenerate our hair cells, curing deafness, imbalance, tinnitus, and vertigo? As usual, we need some background information first, but we will circle back around to these questions by the end of the chapter.

Regeneration is tied to cell growth and repair. In mammalian inner ears, which don't regenerate, all the hair cells are grown in utero. No hair cells develop after birth. But in shark inner ears, which do regenerate, most of the hair cells are not present at birth. Jeff Corwin, a pioneer in hair cell regeneration, estimated that 80 percent of shark hair cells are grown after birth.[1] He studies the macula neglecta, which is a part of the inner ear present in Elasmobranchs, which include sharks, rays, and skates. Humans don't have a macula neglecta. The macula neglecta is located near the posterior canal, and it's a cross between a semicircular canal and an otolith organ. On the one hand, it's shaped like the utricle and saccule, with a bed of hair cells in a gently curving plane. On the other hand, it doesn't have otoconia (ear crystals), like the other otolith organs (the utricle and saccule). Instead, it's overlaid by a gelatinous cupula, like the crista of the semicircular canals.

At birth, sharks have ~24,000 hair cells in the macula neglecta. Considering that the human utricle has ~33,000 hair cells, and the saccule around 18,000, that's a respectable number. But, at birth, sharks are just getting started. The macula continues to grow, adding new hair cells, for life. The adult shark can have a staggering 200,000 or more hair cells in the macula neglecta. This, of course, is why so many people are afraid of sharks! Corwin writes, "In fact, in the gray reef shark, Carcharhinus menisorrah, this macula contains one of the largest hair cell populations described in any vertebrate ear, more than 260,000 hair cells in an individual detector, a number only exceeded by the shark sacculus."

In another study, Corwin examined the reason for the flurry of hair cell growth.[2] This time he studied the thornback ray—*Raja clavata*. Like sharks, the rays also keep growing hair cells in the macula neglecta, starting with five hundred cells at birth, and multiplying to 6,000 cells in the adults (who were seven years old). He found that the hair cells responded to low frequency vibrations, with the best performance between 40 Hz and 200 Hz. So, these hair cells are designed to hear sounds. Interestingly, while the hair cells were multiplying, their connected neurons didn't change in number. With time, each neuron became connected to more hair cells. This suggested that additional hair cells might help the ray hear softer sounds. Corwin confirmed this experimentally, finding that adult rays had better auditory sensitivity than newborns, hearing sounds up to fifty decibels fainter.

Doctor John Carey is a world-famous vestibular expert at the Johns Hopkins Hospital in Baltimore. He is also a personal mentor and friend, as I had the good fortune to spend two years in his fellowship training program. Everyone who meets Dr. Carey and sees him in action comes to the same conclusion: he might just be the kindest and most intelligent doctor you've ever met. In my own practice, when faced with doubt, I'll ask myself: "What would Dr. Carey do?" If the answer isn't obvious. . . . I'll just email him.

In 1996, Dr. Carey and colleagues performed an experiment on chickens.[3] Others, starting in the 1980s, had already shown that after several types of injury, including loud noise and poison, avian inner ears would restore themselves. Even if you killed off every single hair cell inside the cochlea and the vestibule and the semicircular canals of a bird, after a period of time they would magically regrow. Dr. Carey's team wanted to know if the new hair cells—the regenerated hair cells—were functional. Did they work restoring hearing and balance? Or did they just look pretty? To sort it out, he infused the animals with streptomycin. Streptomycin, which is a brother of gentamicin, is a widely used antibiotic. Unfortunately, in addition to its prowess in killing bacteria, streptomycin is also lethal to hair cells. Therefore, in experiments it can be used as a hair cell poison. The animals were then closely observed, and periodically their vestibular systems were assessed. As we saw in chapter 4, one of the best ways to check vestibular function is to assess the vestibulo-ocular reflex (VOR)—the cardinal reflex that keeps eyes steady during movement. Dr. Carey found that for the first several weeks, nothing happened. As expected, after the toxic streptomycin, the chickens had no measurable VOR. It was totally gone, indicating no vestibular function. However, after a week, the VOR started coming back. After two weeks, some chickens started reaching the normal VOR range, and after two months, all the chickens were back to normal. This functional recovery correlated with regrowth of hair cells. Right after the streptomycin injury, almost all the hair cells appeared dead. However, after two months, the research team noted that they couldn't tell the difference between the regenerated cluster of hair cells and their normal appearance. The chickens were so good at rebuilding their vestibular system that you couldn't even tell that they had done so.

Unfortunately, mammals can't regrow hair cells. Once our hair cells die—due to noise, trauma, genetic errors, toxins, infections, or other diseases—they stay dead. But there are so many questions. What allows birds and sharks to regrow hair cells? Why would nature, in all her wisdom, not grant us that ability? Our brains dedicate a lot of real estate toward hearing and balance. If those abilities are so important, and it's biologically possible to endow them with a self-regenerative capability, why didn't evolutionary forces grant us that power? Is there an unknown trade off, with some negative effect that accompanies this miraculous ability? Do we possess all the necessary elements, such that the only barrier is unlocking the genetic recipe that instructs supporting cells to transform into new hair cells? Or is there some fundamental difference between mammalian and nonmammalian hair cells?

HAIR LOSS TREATMENTS

The idea of regenerative medicine may sound like science fiction, but it's not. 350 years after the discovery of the cell, and seventy years after the discovery of DNA, we understand disease like never before. These achievements were transformational. Who would have thought that all animals are a cohesive assemblage of trillions of cells, working together toward shared goals of surviving, eating, and reproducing? Who would have believed that each of those cells contains software—in the form of DNA—controlling fabrication of tens of thousands of microscopic machines (proteins)—like a nanobot 3D printer? Perhaps the most mind-blowing fact of all time: not only are you, with your quirky personality, your detoxifying liver, your childhood memories, composed of an unfathomably intricate arrangement of cells, but your entire body self-assembled from just one cell. It hurts to contemplate just how many things need to go perfectly to allow this miraculous transformation, and yet here we are.

With our modern understanding of the cellular and genetic basis for life, we turn to disease. In some ways, we've made remarkable progress. Well over one hundred specific genetic defects in the cellular machinery necessary for hearing have been proven to cause deafness. Testing is quick and easy; a cheek swab suffices. In other ways, progress is painfully slow. Millions of people die each year from cancer: a process of accumulated genetic mutations that transform our selfless cells into destructive, invasive mutants.

We are now entering into the era of not only understanding how defective proteins and cells cause disease, but how to fix them. The concept of gene therapy is simple: if DNA dysfunction is the culprit behind disease, perhaps DNA repair is the solution. If the cellular software has a bug, then we just need an upgrade to patch the faulty code. For me, the allure of gene therapy is personal. My grandfather died of a genetic disease.

My mom's dad, Jack, passed away when she was six years old. She doesn't remember much of her father, except for a consuming itch that would make him scratch his skin raw. Later, she found out that the intense itch was common in those with uremia, a side effect of accumulated toxins due to kidney failure. Jack was born faulty, with a disease called "Autosomal Dominant Polycystic Kidney Disease." Damage accumulates slowly, and Jack lived happy years as a pharmacist in Seattle, marrying and fathering three girls. His hobby was photography, which I inherited. But all the while, his kidneys were shutting down, and in 1960, his kidneys sputtered out.

Kidneys serve as a waste disposal system for the body, filtering blood to remove harmful substances, which creates urine as a byproduct. When kidneys fail, their function must be supplanted by an artificial kidney, in a process we today call hemodialysis. The first human attempt at hemodialysis was in 1924, in Germany. Blood was extracted from the body, purified, and returned. To prevent the blood from clotting, the pioneering doctors borrowed a trick from nature, instilling leech saliva to keep the blood flowing. The technique was refined over the years, but a critical problem still remained. Repeated puncture of arteries and veins wasn't feasible, because they would quickly scar shut. Because of that, until the 1960s, dialysis was only used as a temporizing treatment—it didn't work for chronic kidney failure. The technology for purifying blood was ready but couldn't be used. A new method of access was needed, whereby the same artery and vein could be used repeatedly for dialysis.

In 1960, Belding Scribner and colleagues at the University of Washington found a solution. They fashioned a shunt made of Teflon and connected it to catheters with silicone elastomer. The shunt was an improvement of a previous glass design and allowed repeated sessions of dialysis. Human experiments began, and the research team dutifully published their results in a series of papers in the early 1960s. The papers describe findings from their first four subjects. In case reports, it's common to refer to each patient by their initials, to preserve privacy. The fourth patient—"JC"—is my grandfather.[4]

Scribner writes, "JC is a 48 year old while married male with a history of polycystic kidney disease." He continues, "by the spring of 1960 he had onset of pruritus, anorexia, and vomiting." The pruritus is the all-consuming itch, burned into my mother's memory. While most of the cohort did well, prolonging life by years and decades, my grandfather did not. The body can only compensate so much for the ravages of kidney failure. Without a mechanism to regulate fluid balance, blood pressure skyrockets, hardening arteries and fatiguing the heart. The heart compensates, building muscle and growing larger, but it's an act of desperation, as the massive heart can't last long. Scribner continues his story in a follow up publication: "three of the original 4 patients are alive after 2 years of treatment with periodic hemodialysis. The 4th patient died after 12 months of treatment. Death occurred suddenly at home while he was sleeping and an autopsy revealed acute hemorrhage into an atheromatous plaque of the right coronary artery." Actually, the original report, in a perhaps a Freudian typo, wrote "plague" instead of "plaque." My mom recalls that the autopsy was a big deal. My family is Jewish, and generally autopsy is discouraged by religious law. Special permission was obtained in this case from the rabbi, in recognition of the scientific benefit. In June of 1961, Dr. Scribner sent a letter of condolence to my grandmother. He wrote, "I am sure you often wonder whether the last year has been worth all the trouble, and maybe in Jack's case it wasn't; but I can assure you that in being one of the pioneers, Jack contributed greatly to our knowledge about this new technique so that others who will follow will suffer less and benefit more."

The success of Scribner's shunt birthed a new problem. Hemodialysis required enormous machines, teams of experts, and considerable time. But it was lifesaving. With limited resources, and countless patients, physicians didn't know how to choose who would live, and who would die. In an act known to all students of bioethics, panels were created to review incoming cases and make the soul-splitting decision of who to save. Today, those with polycystic kidney disease can receive a kidney transplant, trading the horrors of kidney failure for the risks of immunosuppression. But, like all moments in time, this shall pass. In the future, gene correction will certainly be the treatment of choice, providing a quick and lasting cure. The sad tales of little girls, trying so desperately to remember their father, will cease to be.

The most common form of inherited deafness is due to a defect in the gene responsible for a protein called connexin 26. Connexin 26 belongs to a family of proteins called gap junctions that create openings between adjacent cells. Normally, cell membranes form a barrier around the cell, but sometimes it's advantageous to have multiple cells open to each other, similar to opening up those locked doors between hotel rooms to form a suite. In the cochlea, connexin 26 is important for potassium flow. As we saw in chapter 3, potassium gradients are critical for hair cell function, acting like a battery to power the cells. With two defective copies of connexin 26, children are generally born deaf. The gene therapy approach is straightforward: replace the defective gene with a functional copy. The most common issue with connexin 26 is a single typo in the genetic code, with one missing letter (a guanine for those who must know!). Unfortunately, this one error causes the protein-making machinery to turn off prematurely, creating useless, rudimentary fragments of connexin 26. It almost seems too good to be true. Could you use gene therapy to cure the most common cause of inherited deafness, by correcting one tiny typo?

Of course, things are always more complex than they seem. To successfully reprogram genes, several conditions must be met. First, the problem needs to be genetic, or at least fixable with genetic tinkering. You can repair a faulty gene, replace a missing gene, or block a bad gene. As we'll see with attempts at hair cell regeneration, you can even try to use gene therapy to instruct cells to repair damage from other causes. If genes are responsible for building a cochlea or a vestibular system in the first place (during development), then there must be some secret set of instructions that would allow for their recreation. Second, you need to be able to manufacture precise snippets of genetic code. Third, you need an effective delivery mechanism to get the gene therapy code into target cells (without, of course, infecting other cells). Finally, the genetic code patches need to be uploaded and incorporated efficiently into the host cell's DNA.

At the time of writing this book, there are no FDA approved gene therapies for the ear. However, there are a handful of approved gene therapies for other conditions, including a rare disorder where the image sensor at the back of the eye doesn't form properly, spinal muscle atrophy (a terrible, fatal disease where muscles waste away), and several products for advanced-stage cancers. There has been one clinical trial of a gene therapy for hearing loss, and even though it didn't work, it's still a huge step in moving from science fiction to science fact. The targeted gene in that trial is called Atoh1 (sometimes science names are fun! And sometimes they just look like a random alphanumeric glob). Novartis sponsored the trial, which lasted from

2014 to 2019. Twenty-two people with severe hearing loss choose to participate. During the trial, they underwent a surgery to create a tiny opening in the base of the stapes bone (the footplate) and infuse the gene therapy into the inner ear with a microcatheter. Nothing seriously bad happened, but also nothing good. No subjects regained any hearing.

Atoh1, the first ear gene targeted for gene therapy, wasn't defective. Instead, it's a regulatory gene, active during development, that helps control the fate of hair cells. If you get rid of Atoh1, no hair cells develop. Furthermore, if you damage all the hair cells in a mouse, and then administer Atoh1, new hair cells grow. That's why Atoh1 was chosen. It's a powerful master gene of the ear, capable of raising hair cells from the dead. Even though, from a gene therapy perspective, diseases with one defective gene are simpler to fix, the potential market for a hair cell panacea that cures all problems, regardless of cause, is far larger.

I don't think anyone really knows why the Atoh1 gene therapy didn't work. It was an ambitious attempt to recreate the in-utero environment where cellular signals initiate a flurry of hair cell construction. In the ear, we are hampered because we can't biopsy anything. The microscopic ear has no redundancy, and therefore biopsy leads to destruction. From a medical perspective, it's child's play to surgically grab a chunk of liver, kidney, or lung, and send it to the lab for analysis. No change in function is noticeable when you remove 1 percent of an organ. Even the brain has neurons to spare. For hearing and vestibular science, we are forced to rely on animal models and educated guesses. The Atoh1 might not have gotten into the target cells. Or maybe it did, but the DNA payload wasn't integrated. Maybe it was, but the Atoh1 signal alone wasn't enough to regrow hair cells. Adult human ears are not neonatal mouse ears, and it's likely that there are unknown factors in re-sparking long dormant genes. It's also possible that during the study, Atoh1 did cause new hair cell growth, but maybe they were nonfunctional hair cells. Or the wrong type of hair cell. Or there just weren't enough of them. Or it's possible that in these patients, the cause of deafness wasn't just their hair cells, but perhaps another component of the cochlea as well. There's still active research into Atoh1, and a feeling that the story isn't quite finished yet.

For gene therapy to work, you need an effective delivery method. Raw genetic material, like DNA and RNA, rapidly degrades in most environments without a protective shell. To solve the delivery problem, scientists looked to nature. They wanted a tiny delivery truck, capable of both protecting DNA and getting it to target cells efficiently. One type of organism, the smallest *living* things in the world, are incredibly good at those tasks. Viruses are tiny, dwarfed by both bacteria and our cells. They simply consist of encapsulated genetic material. (Author's note: I asterisked *living* because that's up for debate). On the one hand, viruses replicate, infect, adapt. On the other hand, they are parasites, lifeless in isolation, only able to reproduce by hijacking living cellular machinery. Viruses can cause great harm, as is obvious to anyone living through the COVID-19 pandemic. But can they be used for good? Nowadays, almost all gene therapy trials use a virus to get gene therapy into cells. Of course, you have to pick the virus wisely. The ideal virus doesn't cause disease, has sufficient cargo capacity to carry all the gene therapy you want, and selectively only infects certain cells. If we want to cause supporting cells to transform into hair cells, we only want to infect supporting cells, not neurons or other cell types.

To understand the current landscape of regenerative medicine, I interviewed Taha Jan, MD. He's a surgeon-scientist at Vanderbilt University, in Nashville, Tennessee. He explained to me that regenerative medicine, at its simplest, is rebuilding damaged or dead cells with new, functional replacement cells. Currently, there are three ways to achieve regeneration. The first is gene therapy, which we have already discussed. The second is small molecules, as certain small molecules have properties that can change cell behavior, similar to the effects of gene therapy, but without having to directly alter DNA. The third is cell-based therapy, like using stem cells to replenish old, worn out cells.

Dr. Jan said that in the lab, we are further along than most people think. Scientists have been able to take skin cells and fibroblasts (a supportive, structural cell found throughout the body), and by manipulating genes, turn them into stem cells. Adults are like differentiated cells, with highly specialized jobs. One is a video game developer, another is a rock star, and another a professional Lego designer (all my dream jobs . . . sigh!). Children are like stem cells. They haven't yet chosen their profession, so anything is possible. The scientists have then taken these induced stem cells and coaxed them into developing into hair cells. Not only that, but they've built whole organoids—with hair cells, supporting cells, and neurons. "It's an ear—in a dish!." If we can control cells in the lab, converting a fibroblast into a stem cell and then into a hair cell, it would seem possible to do that in people someday as well.

I asked Dr. Jan a question we discussed earlier. If birds and fish and amphibians can all regenerate hair cells, why can't we? He had a few thoughts. First, he explained that cells in the human inner ear are "postmiotic." That means that they are no longer capable of dividing and making more cells. He stated, "As far as I'm aware, there's never been a cancer of the sensory epithelium ever described, which is an interesting fundamental biological question if you think about how these programs have been totally shut down in this epithelium of the inner ear." There is a trade-off here. Dividing cells can be used to repair damage like, say, a broken bone or a skin laceration. But the repair mechanisms are limited, and result in scarring. For bones and skin, that is totally fine. But in the inner ear, scarring may be more harmful than the original injury. Picture a piano repair man who fixes broken keys by gluing together the whole piano. The innate repair mechanisms, designed for traumatic injuries and microbial infections, are counterproductive in highly specialized organs.

Dr. Jan does think that the superspecialization of our mammalian inner ears does have an advantage. We lost the ability to regenerate, but we gained better hearing: "From an auditory perspective, we have a huge advantage because of frequency tuning and the dynamic range that we have for acoustics. You can hear a pin drop to a jackhammer and the orders of magnitude of that dynamic range that really only the mammalian cochlea can provide." Dynamic range is the difference between the lowest pitch sound that you can hear, and the highest pitch sound that you can hear. In humans, the typical range is 20 Hz to 20,000 Hz—20 Hz is a lower than baritone hum, and 20,000 Hz is an ultra-high-pitched squeal. That's pretty good for the animal kingdom, but very average for mammals. Many mammals can hear higher pitches than we can, like beluga whales, who can hear sounds above 120,000 Hz.

In the contest of dynamic range, mammals win by a landslide. A top ocean predator, the wild tuna, can only hear from ~50 to 1,100 Hz. Bullfrogs perform slightly better, hearing from 100 to 3,000 Hz. Even the mighty chicken—king of the coop—can only hear from 125 Hz to 2,000 Hz. This, of course, answers an age-old question. The chicken clearly crossed the road to get closer to the other chickens, because he couldn't hear what they were saying. Hunting at night, a bat (a mammal) can hear sounds up to 110,000 Hertz. The owl (a bird) can only hear up to 12,000 Hz, so it's good that they are so wise, because compared to bats, they can't hear a thing.[5]

A few months before this book was due to the publisher (2024), a major advance was announced in inner ear gene therapy. Several research teams, for the first time ever, had successfully used gene therapy to restore hearing. The corrected gene was otoferlin, which causes congenital hearing loss when mutated. The healthy otoferlin protein is important for communication between the hair cell and its connected hearing nerve. Hair cells talk to neurons with neurotransmitters (e.g., glutamate). Neurotransmitters are chemicals that cause the nerve to fire (depolarize), sending the signal along the brain. To facilitate the process, the hair cell stores packets of neurotransmitters at the interface between hair cell and neuron in to-go bags called synaptic vesicles. That way, as soon as the hair cell hears the sweet melody of "Concerning Hobbits," it can deploy the synaptic vesicles posthaste, wasting no millisecond in relaying Howard Shore's glorious serenade to the auditory cortex. Otoferlin is critical to that process, and without it hair cells struggle to stimulate their neurons.

Otoferlin was thought to be a good target for gene therapy, for a few reasons. One is that the mutation usually causes a unique pattern of dysfunction on hearing tests (preserved otoacoustic emissions with absent auditory brainstem responses). That's an advantage for finding those affected, since most babies with congenital hearing loss don't get genetic testing (although arguably with gene therapy that will change rapidly over the next few years). The other reason is that there was thought to be a good window of opportunity. For gene therapy to be effective, you want a healthy ear that is just missing one part. Some genetic causes of hearing loss cause widespread damage in utero, making it much harder to restore function after birth.

A few teams around the world were all studying otoferlin at the same time. This included a Chinese team, who published remarkable results in *The Lancet* in 2024.[6] They treated six kids with otoferlin gene therapy and were able to restore hearing in five of them. Critically, this included not just the ability to hear softer sounds (five participants), but also the improved ability to understand language (three participants). There were no serious safety concerns, which can happen with gene therapy trials. They used an adeno-associated virus (AAV) as the delivery vehicle to get the otoferlin gene therapy into the cochlea. They overcame a size limitation—the otoferlin gene is larger than the carrying capacity of adeno-associated virus—by splitting the gene into two parts. Two American companies, Decibel Therapeutics and Akouos, are also working with otoferlin gene therapy, and both have announced preliminary, positive results for ongoing clinical trials.

It seems like gene therapy is the future of ear medicine. Our understanding of the genetic basis for hearing loss is more advanced than for vestibular disease, but the two systems are fundamentally more similar than they are different, and restorative treatments for cochlear hair cells may work for vestibular hair cells. Researchers around the world, as well as several biotech companies, are actively working on solutions to each potential pitfall in the process. In the last fifty years, we've managed to decode the secret language of life. I am hopeful that during the next fifty, we'll use that knowledge to heal our balance and ourselves, restoring dignity, happiness, health, and hope.

14

Spacing Out

THE GRAVITY OF THE SITUATION

The vestibular system—the most ancient part of the inner ear—evolved to support life on Earth. Its primal function, shared among nearly all complex species, is to sense gravity. Plants need to grow toward the Sun. Fish need to know where the surface is. And terrestrial creatures must sense gravity to form a mental map of the world, maintain orientation and vision, to safely ambulate and navigate. What happens to us when gravity is no more? What happens to us in space?

Now, I don't mean actual space. That's a quick answer. In the vacuum of space, humans die quickly, from a terrifying combo of air rushing out of the lungs, blood boiling, and freezing/burning, the ratio dependent on surface area exposed to the Sun. Ears would not fare well either. There's an air pocket, in your middle ear, behind the eardrum. This air is going to try to escape, and to do so, it will perforate the eardrum and rush out through the hole. There are no sounds in the vacuum of space because there are no air molecules to vibrate and propagate sound waves. But your eardrums are directly connected to the cochlea, through the hearing bones (ossicles). So in space, theoretically, the last sound that you ever hear would be the sound of your own eardrums exploding.

Aboard a spaceship, thankfully things are different. In a pressurized environment, surrounded by a comfy blanket of air molecules, eardrums are happy. They function properly, picking up ambient sound waves, and passing them along to the cochlea. Unlike airplanes, spaceships are pressurized to sea level. So, as you rocket out of Earth's atmosphere, your ears don't actually pop. Semicircular canals—reliant on inertial forces, but not gravity—also function fine. However, the otolith organs—the utricle and saccule—need gravity to function. Signals are generated by the shifting motion of heavy calcium carbonate crystals atop a bed of hair cells. Without gravity, the crystals just float above the hair cells. The utricle and saccule have lost the ability to function.

Space motion sickness—marked by nausea, projectile vomiting, sensitivity to head movement, malaise, headache, and irritability—occurs to half of all astronauts. Symptoms are felt during gravity transitions: the first few days of spaceflight, and on returning to Earth. Astronaut and former commander of the International Space Station, Colonel Chris Hadfield, describes: "Feeling nauseated is inevitable during the first day or so in space because weightlessness completely confuses your body. Your inner ear no longer has a reliable way of judging up from down, which throws your balance out of whack and makes you feel sick. In the past, some astronauts vomited throughout their entire flights, their bodies just never accepted the absence of gravity."[1]

Evidence points to otolith dysfunction as the cause of space motion sickness.[2] Without the force of gravity pulling inner ear crystals downward, they just hover, suspended in inner ear fluids. Lacking orientation, the inner ear compass doesn't function. There's no earthly correlate of weightlessness, so the brain assumes that bizarre signals coming from the otolith organs indicate generalized malfunction. In nature, sensory distortion is caused by accidentally ingesting harmful substances. To protect the body against the presumed poisoning, the brain instigates a fail-safe mechanism to quickly eject any remnant toxin in the stomach. With space motion sickness, projectile vomiting is often the first symptom.

To understand microgravity effects on the otolithic system, a team of researchers studied cosmonauts before and after a six-month stint aboard the International Space Station (ISS).[3] Constructed of a long central axis of interlocking modules and sprawling arrays of solar panels, the ISS resembles a massive dragonfly the size of a football field. It hurls around Earth in low orbit at incredible speeds, encircling the globe once every ninety minutes. While technically the station is within the domain of Earth's gravity—it's in orbit after all and would eventually crash into Earth if not for periodic rocket boosts—the force of gravity aboard the station is miniscule. The ISS has been continuously occupied since the year 2000, enabling scientific experiments on the long-term effects of space.

To test the otolith system, the researchers studied a reflex known as the ocular counter-roll. The otolith organs, and the utricle in particular, sense tilts of the head, and direct the eyes to roll in the opposite direction of the tilt to compensate. So, if you tilt your head to your right shoulder, your eyes will roll to the left. Unlike the vestibulo-ocular reflex, the gain (the ratio of head movement to eye movement) of the ocular counter-roll isn't one, instead it's estimated around 0.1 to 0.25.[4] So, if you tilt your head fifty degrees, your eyes may only counter-roll by ten degrees. A small change, but enough to be measured. While the tilt movement itself can be sensed by both semicircular canals and the utricle due to the rotational acceleration forces produced, a static tilt can only be sensed by the utricle. To test the ocular counter-roll, cosmonauts were rotated in a centrifuge. They were seated half a meter from the center, the axis of rotation, facing the same direction the centrifuge was spinning at 254 degrees per second. Picture a research-grade merry-go-round. At that speed, the outward centripetal force is equal to downward gravitational force (both 1 g). Therefore, the combined vector of those forces is forty-five degrees. That means that relative to horizontal, the otoconial mass sitting on the utricle will tilt forty-five degrees once steady state is achieved, tilting the world by the same amount. To achieve a static tilt, the investigators waited forty seconds after the start of rotation to begin measurements, eliminating any effects from the semicircular canals, which would have died out by then.

Measurements were done before the mission, within a few days of returning to Earth, and a week later. It would have been nice to do the testing on the ISS as well, but it costs about $10,000 to put a pound of payload into space, and the testing equipment is heavy! Results showed that after spaceflight, almost all cosmonauts had a significant decrease in the ocular counter-roll. This finding helps to prove the theory that microgravity wreaks havoc on the utricle. Thankfully, their responses return to near normal by the second test, done about ten days after landing. The researchers then took things further. Upon returning to Earth, most astronauts also suffer from lightheadedness and low blood pressure upon standing up, called "orthostatic intolerance." On Earth, when you go from lying to sitting to standing, your heart and arteries need to work together to increase pressure to get blood from the heart to the brain. That's because when lying down, your heart and brain are on the same level, and when standing, your brain is two feet above your heart. In space, without gravity, this isn't an issue. Failure of the heart and arteries to increase blood pressure, resulting in low blood pressure in the brain causing lightheadedness, is called orthostatic intolerance. In severe cases, one can even pass out. There are a lot of theories as to why that would be, for example blood pressure regulatory systems may decondition when they aren't used for long periods of time. However, based on research linking otolith function to proper blood pressure control with position changes, the research team wanted to see if changes in the ocular counter-roll correlated with orthostatic intolerance. They found . . . drumroll please . . . that the two were highly correlated. Cosmonauts with larger losses of their ocular counter-roll had corresponding larger losses of their blood pressure when going from lying to standing. In this unique experiment, taking advantage of the microgravity of outer space to incapacitate the utricle, it was proven that the vestibular system helps control blood pressure while standing up. This is a pretty remarkable finding, because many of us doctors were taught that if someone only has dizziness while standing, the vestibular system is not at fault, instead the problem lies with the orthostatic response. Well, using space science, we can see that things aren't quite so simple, and that the vestibular system helps guide the orthostatic response. Other things help as well, like pressure receptors inside our arteries that drive a sympathetic nervous system response, increasing heart rate and constricting blood vessels. But the vestibular system plays a role as well.

In Chapter 7 we explored how the vestibular system is important for thinking. We saw that patients with vestibular disorders complain of "brain fog," the rats without a vestibular system couldn't find their way home, and that vestibular system neurons are connected to brain regions responsible for abstract thought. In space, where dysfunction of the utricle and saccule are guaranteed, does cognition suffer?

Space fog—marked by difficulty focusing, thinking, multitasking, and navigating—is real.[5] The cause of space fog is currently unknown, and to my knowledge no studies have been performed to see if vestibular dysfunction is the culprit. Several other plausible theories have been put forth, such as sleep disruption (no circadian rhythm), radiation, higher carbon dioxide content in the air, and stress. Clearly, more experiments are needed.

THE UPSIDE DOWN

Our monkey brains evolved on Earth. Within any room, we know that the ceiling is up, the floor is down, and that the walls form the sides of the room. When that cardinal arrangement is perturbed, like in an amusement park fun house, we stumble. In a tilted room, we can barely walk. We are so used to visual cues aligning with the dominant and ubiquitous up/down vector of gravity, so our poor brains struggle. Even a slight offset is off-putting, a fact well known to anyone who grew up in an old or poorly constructed house with uneven floors.

But can our terrestrial brains adapt to a world without gravity? In a room in space, no surface is special. The floor and ceiling are just other walls. For the floating astronaut, up and down are relative terms. And from an efficiency standpoint, it makes sense to fill every available surface with equipment. Why keep the floor clear when it's not needed for walking? Why keep the ceiling blank when it's just as easy to reach as the wall? Can we effectively transition from the two-dimensional world of Earth to the three-dimensional world of space?

Charles Oman, a research scientist at the illustrious Massachusetts Institute of Technology, was intrigued by these questions.[6] By combing through astronaut (and cosmonaut) reports and interviews, he was able to categorize several illusions that are routinely experienced in space. During a "visual reorientation illusion," the identity of different sides of the room would suddenly shift to match the astronaut's orientation. For example, astronaut Amanda might enter a laboratory module (space room) and automatically assume that the surface above her head was the ceiling, and the surface below her feet was the floor. However, after getting into position to perform an experiment on the "ceiling." she's feeling a sudden disorientation, as the "ceiling" transformed into a "wall," and a new "ceiling" and "floor" were mentally assigned to the sides of the room above and below her.

A second perceptual trick is called the "inversion illusion." This occurs at the initial transition to weightlessness, once zero gravity is reached. Those affected experience a powerful feeling that they are upside down, and sometimes a feeling that they are flying or somersaulting upside down. Unlike during a visual reorientation illusion, when the environment shifts to align with the astronaut's viewpoint, during an inversion illusion the environment shifts to align exactly opposite gravity. In other words, during a visual reorientation illusion, you always feel right side up, whereas during an inversion illusion, you feel upside down.

Oman's conclusion was that astronauts "show evidence of their terrestrial evolutionary heritage." While there could be advantages—e.g., visual reorientation illusions might allow greater efficiency in spacecraft design since astronauts could learn to work on all six walls of every room, rather than just the four "walls"—there are dangers as well. During an emergency, when astronauts need to quickly move about a space station, disorientation can be deadly. During spacewalks, some describe a sudden shift wherein the Earth goes from being above the spacecraft to below. The sudden feeling that you are barely hanging on to a tiny space bus, speeding past a vast void above the Earth has understandably caused panic and fear. Oman recommends that to prevent disorientation, spacecraft should have a clear and consistent visual design with an up and a down direction. For example, the main lights for each room should be mounted on the "ceiling," and not equally spread on all surfaces. Furthermore, in a station with multiple modules, the "ceiling" in each room should be in the same orientation relative to each other, to prevent disorientations when moving between modules.

Oman's theory was personally confirmed to me the one time that I met an astronaut. That's right, true story, a few years ago we had a former commander of the International Space Station over for burgers. Leroy Chiao was the first Asian American to do a spacewalk, and also the first American to vote in space. He had also flown several missions on the space shuttle, a fact that became awkwardly apparent later in the evening. Trying to impress him, I took him on a tour of my office and proudly presented a LEGO model of the lunar lander, and the Saturn V rocket that carried it into space. "You know, Jeff," he said, "I think LEGO also makes the space shuttle."

Halfway through dinner, I couldn't resist anymore. I interrupted the conversation and inquired about vestibular dysfunction and disorientation in space. Commander Chiao had spent 229 days in space, and thirty-six hours of "extravehicular activity." He reported that both were quite common, and that once he had discussed with NASA an idea to paint arrows on the outside of the ISS, pointing toward the airlock, to help guide the astronauts to get back inside the station after a spacewalk. In other words, in space, you can get lost walking around your own block.

SPACE JELLYFISH

In the fantasy universe of *Star Wars*, it's easy to planet hop. With a hyperdrive, you could ski on Hoth, watch a double sunset on Tatooine, and then go clubbing on Coruscant. Sadly, in our reality, the laws of physics dictate a speed limit—the speed of light. While fast on the scale of Earth, or our immediate vicinity (sunlight takes eight minutes to reach Earth), on the scale of galaxies, it's really slow. Apollo astronauts reached the moon in three days. Several months of travel are required to reach Mars. It took the New Horizons probe nine years to rendezvous with Pluto. And that's all within our corner of the universe, the solar system. The closest star to our Sun, Alpha Centauri, is four light-years away. At the time of writing, the fastest spaceship is the Parker Solar Probe, which will reach a top speed of 430,000 miles/hour in 2024, during its final approach to the Sun. At that phenomenal speed, it would take about 6,700 years (!) to arrive at our nearest star. Even if speeds are dramatically increased in the future, and I'm sure they will be, it will still take years to travel the cosmos. It won't just be enough for humans to survive for years in the void. For intergalactic travel, humans are going to have to be born in space.

How will human development be affected by the gravity-free environment of space? Well, no one knows for sure, as to date, no human has gestated in space. In fact, according to NASA, no one has had sex in space (or at least no one has fessed up!). Now, there will likely be a veritable cornucopia of issues for any child born in space (muscles, bone, heart, etc.). But this is a book on the vestibular system, so we'll focus on that.

Aurelia aurita, or moon jellyfish, are found throughout Earth's oceans. Their translucent bodies are shaped like a bell, and they move with coordinated undulations called "pulsing." In an aquarium, a mass of jellyfish pulsing, with only the edges of their bodies catching enough light to be seen, can appear beautiful, like an underwater ballet. In the wild, a swarm of jellyfish is terrifying (at least to me, maybe you are braver?). Jellyfish don't have a brain, or complex sensory organs. However, they do have a rudimentary vestibular system, which senses gravity. Located in small organs called rhopalia, are statocysts. Statocysts work similarly to our otolith organs—heavy crystals (statoliths) are present within a sphere lined internally by hair cells. Since the crystals always fall downward, the animal can tell which way is up. This information helps guide pulsing behavior in the jellyfish, telling them where to swim.

Adult moon jellies measure about a foot across, but larval jellies, ephyra, are tiny. So small, in fact, that hundreds of them were packed in seawater, and launched into space as part of the NASA space life science mission in 1991.[7] The mission was nine days long, and the timing was chosen so that the ephyra would develop their vestibular organs while in space. When they got back to Earth, the jellyfish sensory organs underwent a detailed microscopic evaluation. In addition, pulsing behavior was observed.

Dorothy Spangenberg and colleagues found that the space-grown rhopalia and statoliths looked pretty similar to Earth-grown versions. In other words, they weren't grossly misshapen when developed in space. However, in space the jellies developed more statoliths per rhopalia (more crystals in each vestibular organ) than on Earth. This finding was specific to animals who developed in space. Those who were a little older, and had developed on Earth but, courtesy of the space shuttle, took a vacation in space, had normal numbers of statoliths. However, about half of the space statoliths underwent demineralization. Finally, the investigators observed that about 1/5 of the jellyfish who developed in space had abnormal pulsing. They wrote, "pulsing defects in the space-developed ephyrae included uncoordinated pulsing, after-twitches, spasms, and arms out of synchronization during pulsing." Many jellyfish born in space don't develop normal vestibular systems, which affect their ability to perform their most basic function—swimming.

Mammals seem to suffer from similar issues. April Ronca has a PhD in neuroscience and has spent her career studying the effects of microgravity on development. She is the lead author of a study in which pregnant rats were launched into space.[8] A typical rat pregnancy lasts just three weeks, and these rats spent the last two of those weeks in space. The experiment was timed so that the mother rats would give birth shortly after landing, and the research team could study the animals during the first few days of life. The animals were compared to a group of matched-age rats who were put into the same habitats at the exact same time; only the control rats stayed on Earth at the Johnson Space Center. Ronca found a number of abnormalities in the rats who had gestated in space. First, their righting reflex—the basic ability of rats to get on all fours after being placed on their backs—was impaired. The righting reflex test was done in water, which is a more selective test for vestibular function than testing the rats on a dry surface, because they can't use other sensory clues (proprioceptive and somatosensory) in water. Furthermore, they then studied the rat's brains and found that poor performance with the righting reflex was correlated with an immature appearance of neurons going from the gravity sensor in the inner ear (the saccule) to the vestibular nucleus in the brainstem. Compared with normal neurons, those that developed in space had fewer branches. So, not only was behavior affected when rats were grown in space, their brains showed measurable differences in the health and maturation of gravity sensing neurons.

In a separate experiment, two-week-old rats were sent into space for two weeks to find out how microgravity affects this critical period of postnatal development.[9] When they got back home, the righting reflex was studied. Compared to fellow rats who had stayed on Earth, the space rats were again bad at righting themselves. Taken together, these experiments argue that mammals need gravity during prenatal and postnatal development for optimal vestibular function. Will humans who develop in microgravity not be able to stand up if they fall? For starships to safely traverse the vast distances of space, it seems like we are going to need some form of artificial gravity.

Upon returning to Earth's gravity, astronauts experience dizziness, disorientation, and motion sensitivity. Gravity feels super powerful. Chris Hadfield describes his experience, "Back on Earth, though, gravity was suddenly pulling me down and the floor was holding me up, trapping my inner ear in what felt like a constant acceleration that, inexplicably, my eyes couldn't perceive. It's extremely nauseating, worse than the most sickening ride at the fair."[10]

Thankfully, though, it's the gravity transitions that produce symptoms. Once astronauts acclimatize to microgravity, they function pretty well. In fact, a curious thing happens. They eventually become more resistant to motion sickness. British astronaut Tim Peake conducted a fun self-experiment, which is archived on YouTube. He rolled himself into a ball and had colleagues spin him around as quickly as possible. Twirling around rapidly in the middle of a laboratory module, Tim calmly describes that he does not feel motion sick at all. It's clear to any viewer that rotating at that speed on Earth would be extremely nauseating. Why is that?

One theory is based on vestibular compensation. Whenever the vestibular system faces a problem, several brain mechanisms kick in to help dampen faulty signals. In addition, sensory reweighting places increased reliance on visual cues rather than vestibular ones. There is some experimental evidence for this theory.[11] The gain of the vestibulo-ocular reflex in space appears normal, indicating normal semicircular canal function.

However, as we saw in chapter 5, when spun around at a constant velocity, nystagmus will appear due to the initial acceleration and then decay in an exponential fashion. The time necessary for decay (technically for 67 percent of the decay) is called the time constant. It's been long recognized that the time constant is longer than expected from the mechanical viscoelastic properties of the cupula, reflecting a central process that serves to enhance the vestibular response. In other words, velocity storage is an amplifier for the vestibular signal. The time constant reflects the signal from the semicircular canal, plus the velocity storage. At 0 g, the time constant is markedly decreased, compared to testing at 1 g. This may reflect the brain's attempt to reduce the vestibular signal, by turning off the velocity storage mechanism. So, in microgravity, when the vestibular system is wonky, our brains have a method of diminishing its influence. With the vestibular system in restraint, it makes sense that astronauts would be more resistant to motion sickness.

Outer space is a remarkable place. Besides the views—which are out of this world (ha!)—it's a unique environment in many ways. As we saw, without gravity, the otolith organs don't really function, which has profound implications beyond the initial bout of space motion sickness. The ability to think abstractly, develop vestibular organs in utero, and coordinate complex antigravity movements may all be affected. If we ever want to travel the stars, we'll have to solve these problems.

PART 5

THE ENDING

15

Spinning Out of Control

Advice for Patients

finished this book, and then realized it wasn't complete. I set out to write a fun book, with stories and ideas that hopefully bring the vestibular system to life. But, of course, far too many suffer from vestibular problems. This book isn't a self-help book. But, with that said, my patients motivate me to do everything I do, and I can't have a book without some practical advice. So, this section is optional. If you've never suffered from vertigo or dizziness or imbalance, count yourself lucky, and get outta here! But if you are beleaguered by symptoms, then this chapter is for you.

A legal disclaimer: Medical care is of course provided through interactions with medical personnel. I am not engaging in a therapeutic relationship with you. I am providing general advice, which you can then discuss with your doctor.

There are a few key points that I want to cover:

- How to find vestibular specialists
- How to prepare for visits
- Thinking about possible diagnoses
- Testing
- The keys to treatment success
- Resources

HOW TO FIND A VESTIBULAR SPECIALIST

In general, there are two types of doctors who treat vestibular disorders: Neurologists and Otolaryngologists (also called Ear, Nose, and Throat Doctors or ENTs). Those are both specialties of medicine. Within each, there is a subspeciality that focuses on vestibular disorders. Within neurology, it's usually called otoneurology. Within ENT, it's usually called neurotology. I am a neurotologist. I have very close colleagues in otoneurology. The main difference is that I do surgery, and they don't. Another difference is that neurologists understand the brain and nervous system (generally speaking) on a better level than neurotologists. Here is the major thing to know. Most neurologists, most ENT doctors, most primary care doctors, and most emergency room doctors (at least in my opinion, forgive me) don't truly understand the vestibular system. Even among neurotologists, many choose to focus on hearing problems and surgical diseases. That means that if you don't do research ahead of time, and figure out who the right person is, it's possible that you won't end up seeing a true expert. Now, that may be ok depending on the complexity of your problem, but it's important to know the landscape so that if you have a difficult-to-treat problem, you don't give up prematurely.

There aren't enough vestibular specialists in the United States and the rest of the world. There are a couple reasons for that. One reason is that otolaryngologists, because of their training (and also for reimbursement reasons) are interested in doing surgery, and most vestibular patients don't need surgery. There are also too few vestibular fellowships (in both neurology and ENT). So how do you know who to see?

In general, for emergencies you should go to the emergency room, and for things that can wait, you should start with your primary care doctor. For vestibular problems, you *need* a primary care doctor. In some areas of the country, it can be challenging to get one, but it's a necessity for good care. ER doctors excel at making sure you don't have a life-threatening condition, like a stroke. They are not trained to diagnose nonemergency vestibular problems. Primary care doctors excel at treating you as a whole person. Remember that an ENT doctor is only going to look at your ears. If you have a problem affecting your whole body, they might be too zoomed in, and they'll miss it. Primary care doctors generally take a holistic approach, looking at overall health, examining blood pressure, the heart, the lungs, the kidneys, the thyroid, electrolytes, and the liver. Dizziness is a nonspecific symptom, and frequently it occurs because of a problem that is easy for a primary care doctor to identify and treat.

The best way to find a vestibular specialist is to do a background check on physicians ahead of time. Most physicians have a web presence, which lists training and interests. Pay special attention to residency, fellowship, and listed areas of expertise. If you aren't making progress with local doctors, then consider seeking out an opinion or treatment from a university.

Physical therapists are invaluable in treating vestibular diseases. But just like physicians, not all specialize in vestibular diseases. You have to check with each office ahead of time. There is also an incomplete list of vestibular-minded physical therapists at neuropt.org. In general, a vestibular physical therapist should be skilled at treating BPPV (benign paroxysmal positional vertigo/loose crystals), unilateral vestibular loss, and bilateral vestibular loss. In addition, an advanced practice might offer treatments for PPPD (persistent postural perceptual dizziness), vestibular migraine. Finally—and critically—physical therapists are essential to preventing falls. They can work with you on strength, balance, posture, gait, and ways to recover from falls. They can also help recommend devices like canes, wheelchairs, and walkers, and they can help with home safety to prevent falls. In the care of vestibular patients, I rely heavily on my physical therapy colleagues.

HOW TO PREPARE FOR VISITS

Here is the truth—whether or not we want to admit it. Doctor time is limited! Use it wisely. If you are scheduled for a fifteen-minute block, and you spend twenty minutes trying to explain your symptoms, your doctor is going to feel rushed. By the end, they are just going to try to say anything to help move the visit along, because they have other patients to get to. You will both end up being frustrated, and you won't get the help you need. So then, how do you make the most of doctor time?

There are a few things you can do to be helpful. The first is to keep really good records. The second is to have all the information available to you. Many patients believe that all the information is "in their file." Unfortunately, the reality (and any doctor will tell you this), is that medical files are a giant mess, and most doctors don't have the time to look through them carefully. An average-sized medical file can be hundreds or thousands of pages, and most of the content is auto-populated garbage aimed at increasing billing. You need to have a one-page summary printed out, for every visit. There should be a one-paragraph summary of the reason for the visit. The questions that you want answered should be written out. You should list every doctor you have seen for this problem, and every test done, every treatment done, and the results. You also need all your medical conditions listed, your medications (including dosage), and prior surgeries. You will get so much more out of your visit if you do your homework. And the reality is if you don't do this work, no one else will.

Another helpful thing is to let the doctor interrupt you. In some ways, being a physician is like being a skilled interrogator. You know how to quickly get the information that you need. If someone doesn't provide direct answers, or goes off on a tangent, then you may need to interrupt. Now, there is a flip side to that. If the doctor is interrupting just to get through the visit quickly, without taking an interest in your case to figure out exactly what is going on, that is a problem. Try to answer the question being asked. Estimates are ok. For example, if you are asked how many vertigo attacks you have in a month—give a number! If you are asked how long they last, give a time estimate. It's not important if your attack was eighteen minutes or thirty-five minutes. It *is* important if it was eighteen minutes or three days. Try your best not to go on tangents or tell stories. If doctors had unlimited time, they would love to hear everything. But they don't.

Be especially cautious about records from another facility. Don't trust anyone who says that they "sent the records over." Every single clinic, I see patients with incomplete records. Get a copy of all your notes and test results, print them out, and keep them in a binder that you bring to every visit. Don't overwhelm your doctor but have everything available just in case. Also, for any imaging studies (MRI or CT scan), bring a CD with the images. Most radiology storage systems don't talk to each other (which is obviously stupid, but that's the reality), so unless you bring with you actual images (the type of file is called a DICOM) on a disk, there is a chance that the doctor won't be able to review them. So, do NOT rely on others. Keep all your records, so that there is nothing that the doctor could ask for that you don't have. Getting someone's records after the visit isn't as effective. Doctors are very busy and it's impossible for them to remember the thousands of patients that they see every year. The visit is your time for undivided attention. Assume that anytime a doctor is reviewing something outside of an office visit, it's divided attention.

Record your visit. Ask for permission. If you can do an audio recording, then great. If not, then take notes. Human memory is *way* worse than we think it is. You will likely want to review things after the visit.

Final advice: bribe your doctor! You want him/her on your side. We are all human. If financially feasible, bring a small gift with you to the visit, like cupcakes for the office staff. Get them at Safeway, on sale, for $4.99. Make sure they are colorful. It's the thought that counts. Doctors spend their entire lives trying to help people, and validation of their efforts feels amazing to them. Obviously, they need to do an excellent job for you, no matter what. Of course. But, we aren't robots, and I personally think that someone will go the extra mile for you if they know that you appreciate them.

THINKING ABOUT DIAGNOSES

Medical care works best when it's a partnership between doctor and patient. So, the more knowledgeable that you are, the easier it is to get the help you need. I created a website with short, educational videos about different vestibular problems.[1] It's a good place to start.

The first thing to do is to figure out what your symptoms are. Make a list. *Vertigo* is the false sensation of movement. That could be spinning, rocking, being pushed, swaying. *Dizziness* is spatial discomfort without vertigo. That means that you feel uncomfortable about where you are in space and how you move through space. *Imbalance* is the feeling of having bad balance, like you are going to fall. *Syncope* is passing out. The vestibular system doesn't cause syncope, so if you are passing out, then you need to talk to your primary care doctor about other causes for your symptoms.

It's good to note triggers and associated symptoms. Make a list. Those can be important clues as to what is going on. Here's an example of a patient who is making things easy for her doctor: "Hi! I am thirty-four, and I only take medications for allergies. For the last two years, about once a month, I get a spinning feeling lasting hours. Afterward, I usually get a headache." This patient is giving specific information on frequency and timing and associated symptoms. All really helpful. Here is an example of a patient who is making things hard for his doctor: "Hey Doc! You have to help me. Something is wrong. I was at the store, and Sally was there, and we were talking, and then it hit me, and we went to the ER, and the doctors there couldn't figure it out. They gave me some medications, which I think helped a little bit. They did a head scan, but I don't know what the results were." While this patient spoke longer than the first patient, they provided less information. Be specific and clear. Make things as easy as possible for your doctor!

Most commonly, vestibular disorders will cause dizziness and/or vertigo. As we saw in chapter 11, superior canal dehiscence causes its own set of symptoms, including hearing your eyes move, hearing your heartbeat, and dizziness with loud sounds. And, as we saw in chapter 10, bilateral vestibular loss doesn't cause vertigo. It causes unsteadiness, and a blurring of the world with head movement. If you have dizziness/vertigo, the first question is whether it's a onetime thing, or a repeating event. If it just happens once (either with full recovery or with some residual unsteadiness), then you have to consider vestibular neuritis or a stroke. If you have recurrent episodes, then you have to consider BPPV (loose crystals), Ménière's disease, and vestibular migraine.

BPPV (see chapter 8) is very common. However, you do need to be aware that there is a misconception out there that it's the *only* cause of vertigo. That isn't true. BPPV vertigo is brief, lasting seconds, up to a minute. BPPV is *always* triggered by changes in head position, like rolling over in bed. So, if you have twenty seconds of vertigo triggered by moving your head, then it's likely you have BPPV. If that is the case, then you should be fine seeing most ENT doctors, most neurologists, and any vestibular physical therapist. They should be able to diagnose you by putting you into the Dix-Hallpike position (leaning back with your head turned to the side), and treating you with an Epley maneuver, which has a very high success rate. If your diagnosis is confirmed, and you get fixed with treatment, then you can even treat yourself at home if it comes back.

Vestibular migraine is *very* common as well. If you have a history of migraine headaches, and you have episodes of dizziness/vertigo, then it's likely vestibular migraine. Keep an eye out for migraine symptoms during attacks, especially a sensitivity to lights (e.g., needing to turn off lights), and sounds (e.g., preferring quiet). Vestibular migraines can feel very different than a migraine headache, don't let that throw you off.

Ménière's disease is marked by hearing loss, vertigo attacks lasting hours, tinnitus, and ear pressure. In most cases of Ménière's, only one ear is involved. So, if you can't point to the ear that is the problem, then it's not likely to be Ménière's disease. A hearing test is essential for determining if it's Ménière's disease. Therefore, it's useful to see if tinnitus and ear pressure accompany vertigo attacks, because they are helpful clues.

TESTING

There are a couple of standard tests used in the workup of dizziness/vertigo. No test is perfect, and no test can figure out what is going on 100 percent of the time. No test is better than a good history and physical with an expert doctor. However, they can provide specific pieces of information that can be really helpful.

An audiogram is a hearing test. I know it seems counterintuitive, but a hearing test is actually really helpful for vestibular disorders, basically because many diseases affect both. A hearing test is useful for Ménière's disease, labyrinthitis, superior canal dehiscence, and vestibular schwannoma. It's also noninvasive, and there is little downside. Numerous studies have shown that people are not good at estimating their hearing. Therefore, almost always, you will want a hearing test. The main exception is BPPV, in which case a hearing test just isn't necessary.

Two types of images are used for vestibular disorders: CT (computerized tomography) and MRI (magnetic resonance imaging). CT can be done with or without contrast. Generally, contrast isn't needed. CT sees bone really well, so it's the test of choice for looking at bone. There is a difference between head CT, which is a rough look generally done for stroke or head trauma, and a temporal bone CT, which is a high-resolution look at the ear. A head CT cannot see the ear well, so it isn't helpful for any ear diseases. The high-resolution temporal bone CT is necessary in the workup of superior canal dehiscence, to see if bone is missing over the superior canal. MRI scans still seem like magic to me. They provide a great view of soft tissue, which includes the brain. Therefore, MRIs are really helpful for looking for stroke, brain tumors, and some neurologic diseases like multiple sclerosis. Unlike a CT, no radiation is used with an MRI. They do take quite a bit longer, and many patients do get claustrophobic in the narrow MRI tube. I personally have. They are also noisy. Both types of imaging—CT and MRI—cannot "see" most causes of dizziness/vertigo, so keep in mind that most of the time, they will be normal.

Vestibular testing is used to check the health of the vestibular responses. You can stimulate the vestibular system a few ways, with temperature (calorics), body movement (rotary chair), head movement (head impulse testing), and sound (VEMP—Vestibular Evoked Myogenic Potentials). It's important to know that vestibular testing can be uncomfortable (after all, you are trying to probe the vestibular system by causing vertigo), and therefore it may exacerbate symptoms. You should get a ride home after and ideally take off a little time to recover.

There are several tests to choose from, and the choice of test really depends on what the facility has—equipment is expensive—and what question you are trying to answer. Caloric testing involves putting hot/cold air/water in your ear canals to cause nystagmus and then measuring eye speeds with infrared goggles (so you can be seen, but cannot see) and a computer. Rotary chair testing is spinning you around on a mechanical chair attached to a motor, recording nystagmus again with infrared goggles. With video head impulse testing, a high-speed camera measures eye position and head position during a head impulse, which is a quick head twist designed to elicit the vestibulo-ocular reflex. All of the above tests are useful for looking at vestibular loss, either in one ear (unilateral) or in both ears (bilateral). The caloric test and the rotary chair test only test one of the three semicircular canals—the horizontal canal—while the video head impulse can test all three canals. In addition, caloric testing frequently involves checking other eye movements, so it can be helpful in diagnosing central (brain) problems causing nystagmus or dizziness.

VEMP (vestibular evoked myogenic potential) testing is different. It seems more circuitous. You use sounds to stimulate the utricle or saccule (with loud enough sounds, both organs can "hear"), but the main use of the test isn't to check the utricle or saccule, instead it's to diagnose superior canal dehiscence, which causes a characteristic increased response. So, if you have symptoms of superior canal dehiscence (see chapter 11), then getting a VEMP test is really helpful. There are two flavors of VEMP—ocular and cervical. They differ in how they measure the evoked response, from one of the eye muscles (the inferior oblique) or from a neck muscle (the sternocleidomastoid). The eye VEMP tests the utricle, the neck VEMP tests the saccule. The best way to think about VEMP, at least for now, is as a superior canal dehiscence test.

You may have noticed that there isn't a test for several common vestibular conditions, like vestibular migraine or PPPD. Those diseases are diagnosed by criteria—based on a careful history and physical exam—and not based on tests. So, really there is no substitute for a doctor doing a good job and being a good vestibular detective.

THE KEYS TO TREATMENT SUCCESS

The main advice I would give is to be a partner in care. Keep good records and follow up on all recommendations. There is a natural tendency to discount treatments that are not medications or surgery. However, vestibular physical therapy, lifestyle changes for vestibular migraine, or cognitive behavioral therapy for PPPD can be highly effective. If you are given medications, keep track of the name, dosage, how long you took it, side effects, and whether or not it was effective. A daily diary can be really helpful, especially if you are having frequent symptoms. Frequently, when we are trying to treat a vestibular disorder, the devil is in the details.

In general, a symptomatic medication, like zofran (antinausea) or meclizine (anti-vertigo), should only be taken when you have symptoms. Preventative medications, like nortriptyline for vestibular migraine, should be taken every day. To assess if a medication is effective, you need to take it for at least three months. Many vestibular patients are medication sensitive, so it's better to start with low doses, and taper the medication on and off. If you do experience a side effect, you need to determine if it's a minor annoyance—in which case you can keep taking the medication—or if it's a serious issue—in which case you might need to stop the medication.

There are many treatments for Ménière's disease. I wrote a patient guide to Ménière's, to help people navigate all the options, and decide what treatments are best for them. It's freely available online.[2]

In my experience, most vestibular diseases can be treated, and most patients will improve with treatment. Some diseases are definitely harder to treat than others. I've personally had a handful of patients with PPPD, chronic vestibular migraine, bilateral vestibular loss, bilateral Ménière's disease, post-traumatic dizziness, and some central causes of dizziness who—despite my best efforts and intentions—I don't think I was able to help. Thankfully, those patients are the exception, and there are some newer therapeutics on the horizon that may help each of them.

SELECTED RESOURCES

- *UCSF Balance and Falls Center* (www.ohns.ucsf.edu/balance-falls). This is my website, and we have professionally produced "vestibular videos" that explain vestibular diseases and concepts in short, fun clips.[3]

- *The Vestibular Disorders Association* (VEDA, www.vestibular.org). Every vestibular patient should know about this organization. The website is very educational, and there are support groups, newsletters, events, and many other resources.[4]
- *Dr. Tim Hain's website* (https://www.dizziness-and-balance.com/). This site has a ton of well-researched information. It's written more for physicians and those familiar with vestibular concepts, but it's still a great resource for everyone.[5]
- *House Ear Institute/House Institute Foundation* (https://hifla.org). The world-famous House Ear Institute in Los Angeles has a YouTube channel for expert lectures. Many of these cover vestibular topics, and they are from some of the most respected doctors and researchers in the field.[6]
- *Association of Migraine Disorders* (https://www.migrainedisorders .org/). This is a very helpful resource for those with vestibular migraine and other migraine problems.[7]
- *Bárány Society.* https://www.thebaranysociety.org/. International group of vestibular-minded folks. Of note, they publish diagnostic guidelines for each vestibular disease, available on their website.[8]

Epilogue

Vestibular Dreaming

Years have passed since I wrote *The Great Balancing Act*. Book sales were slow at first. The story, if it could even be called that, took place outside the Marvel Cinematic Universe, and was therefore ignored. Vestibular scientists tried to spread the word, but even counting friends (few) and spouses (fewer), their numbers were too paltry to matter. I had unwittingly birthed a new genre: "unpopular science." But humans, in their quest for knowledge, could not be held down for long.

In the spring of 2035, that all changed. A prominent public figure, Dwayne Johnson, took ill with vertigo. The "Rock" had developed a case of loose crystals. When the vestibular system attacks, Dwayne Johnson fights back. Ravenous to understand the inner ear, he searched and searched, until an old, dusty volume of *The Great Balancing Act* was discovered, in a collapsed section of the Library of Congress, under a tattered copy of *The Brussel Sprouts Cookbook: A Culinary Adventure.* Determined to learn, he ignored the slapdash writing and inane examples, devouring the book in one, marathon session. Wanting to see the scourge of vestibular disease eradicated from the face of the Earth, he then directed his followers to read it as well.

A quiet revolution began. In dive bars, hotel pools, hiking trails, and airport bathrooms, people were talking. Hushed whispers of a secret knowledge. A forgotten part of the human body. A part of the ear that wasn't for hearing. How could it be? Are small and precious things even real?

Naysayers abounded. "We've seen the *Sixth Sense*," they said, "and it clearly refers to the ability to see ghosts." "Modern science cannot explain hauntings, phantasms, or demonic possession, and is therefore useless." The Pentagon was briefly intrigued and even began a program to genetically manufacture soldiers with supersized semicircular canals. These warriors, imbued with the agility of a cheetah, would be able to accurately fire automatic weapons while flying a jet pack. However, the program was scrapped when the human soldiers' program was terminated.

The Great Balancing Act began to climb in book rankings. At first, there were little surprises, as the "political tell-all book exposing corruption written in retaliation for being fired" genre was considered overdone. However, heads began to turn when *The Great Balancing Act* cut into the very lucrative "self-help book where it turns out your diet is responsible for your terrible relationship with your parents" share of the market. After a long climb, *The Great Balancing Act* unseated the king of publishing— "vampire stories where a teenage girl falls for a centuries old undead creature who has a heart of gold; only eating side characters."

But then, something unexpected happened. A fundamental shift in the great balance of humanity. With new readership, science-minded humans were now in the majority, for the first time in history. "If we can understand the vestibular system," they reasoned, "surely we can fix humanity." It seemed like most political leaders had failed for a simple reason. They had ignored human psychology. The new humanity, recognizing at the same time man's incredible capacity for both good and evil, designed systems of rule to exploit greed when it led to the greater good, and to punish greed when it led to the ruin of many. Self-serving tyrants, who enriched themselves at the expense of their citizens, were eliminated. Bitter conflicts, based on old hatred, were set aside. The simple fact was the three semicircular canals are identical across the human species, meaning that we are all brothers and sisters, and must love each other. Health care became a human right. Racism was eliminated, because if I have otoconia, and you have otoconia, are we not the same?

A golden age began. We realized that if a human starves to death, the vestibular system dies as well. Famine was eliminated. The vestibular system was made for life on Earth and didn't work in space. Therefore, we had to protect our mother planet, keeping it safe for future generations. But vestibular secrets remained. World governments poured money into scientific discovery, with a pledge to double scientific funding every five years. The hiccups were cured, then cancer. Once aging was solved, the true value of human life became apparent, and murder ceased to be. With immortality, riches, and a renewed planet, humanity focused on what was important—acquiring as many dogs as possible.

Thanks for spending precious time with me and my thoughts.

Illustrated Glossary

ORGAN. A specialized part of the body, with a specific purpose, like pumping blood (heart), nutrient extraction (intestines), or vision (eyes).

TEMPORAL BONE. The part of the skull that contains the organs of hearing and balance, located on either side of the skull.

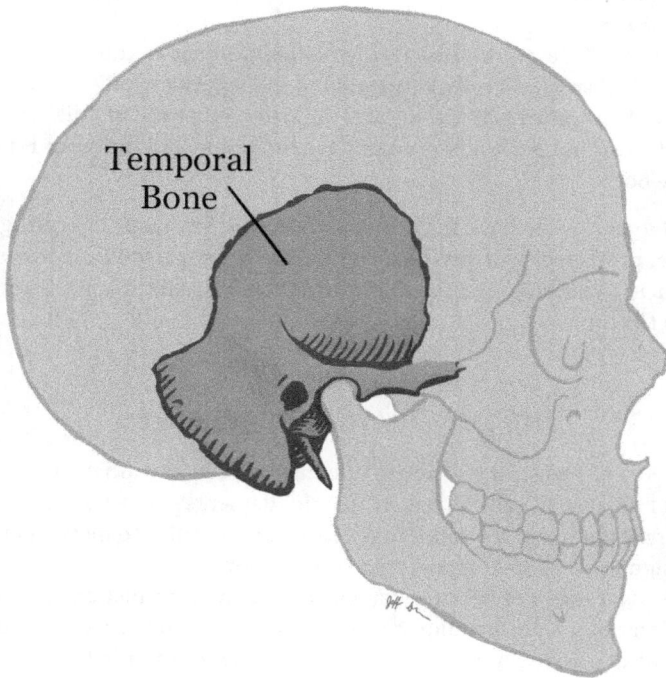

FIGURE 15.1 The temporal bone.

THE OUTER, MIDDLE, AND INNER EAR

THE OUTER EAR. The visible part of the ear, made of a cartilage skeleton, enveloped in skin, and including the ear canal.

THE MIDDLE EAR. The land of the ear bones (ossicles). The space beyond the eardrum, where sound waves are transferred along the chain of ossicles to the inner ear. It's filled with air, replenished periodically through the Eustachian tube. Connected to the air cells of the mastoid.

- TYMPANIC MEMBRANE. The eardrum.
- OSSICLES. The tiny bones of hearing.
- MALLEUS. The first hearing bone, enmeshed in the eardrum, connected to the incus.
- INCUS. The second hearing bone, between the malleus and the stapes.
- STAPES. The final hearing bone, between the incus and the vestibule, the gateway to the inner ear. On one side of the stapes footplate lies the air of the middle ear, on the other side lies the fluid of the vestibule.
- EUSTACHIAN TUBE. A tunnel, partly of cartilage, partly of bone. It connects the back of the nose (nasopharynx) to the middle ear. Normally closed, it opens with swallows and yawns to equalize the air pressure in the sealed-off middle ear.
- MASTOID. The mastoid is mostly behind and above the middle ear. It's a network of small air cells, connected to the largest air cell—the ANTRUM—which then connects to the ATTIC, the space above the middle ear. Since the middle ear and mastoid are connected, infection and fluid can rapidly consume both.

THE INNER EAR. Where the magic happens. Here, the organs of hearing and balance lie, protected by dense bone. There are three parts to the inner ear: the cochlea, the vestibule, and the semicircular canals. All three parts rely on hair cells to function.

THE MEMBRANOUS LABYRINTH

OTIC CAPSULE. The strong bone capsule that protects the inner ear.

HAIR CELL. A special cell, with microscopic filaments (STEREOCILIA) that protrude from the top of the cell. These "hairs" are sensitive to movement, sending nerve signals when bent in a particular direction.

COCHLEA. The organ of hearing, capable of reading information in transmitted fluid vibrations, transforming that signal by volume and pitch, and stimulating the cochlear nerve, which carries the decoded sounds to the brain.

OVAL WINDOW. An oval shaped hole in the otic capsule bone. The stapes sits on the oval window, suspended in place with the annular ligament.

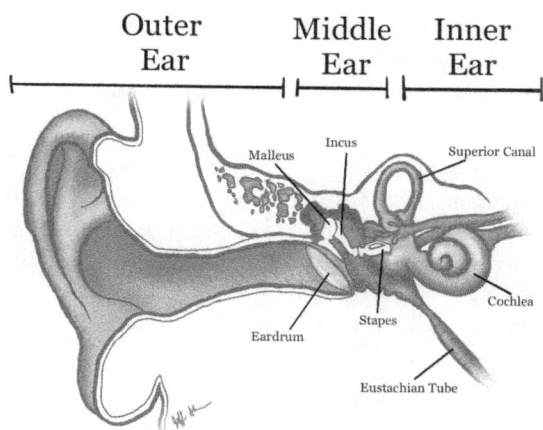

FIGURE 15.2 Outer, middle, and inner ear.

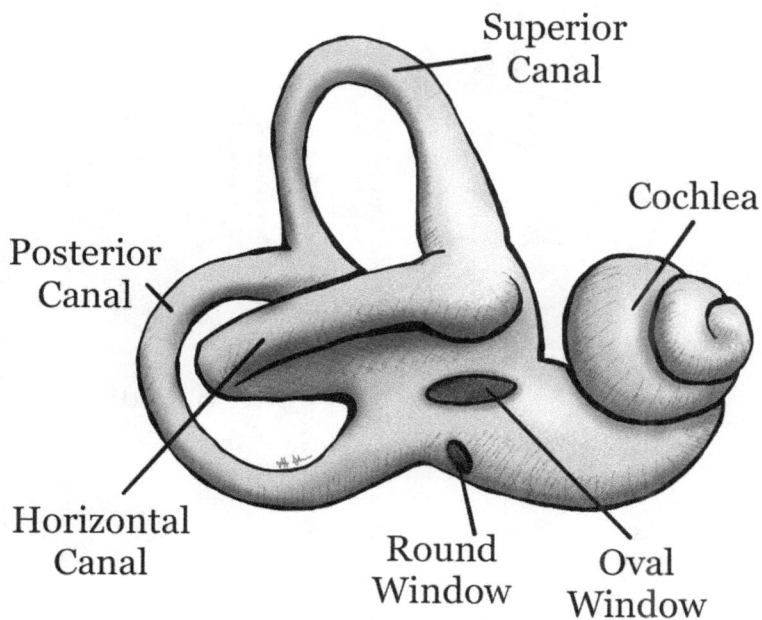

FIGURE 15.3 The inner ear.

ROUND WINDOW. The second opening in the otic capsule bone. It's covered by a membrane, and it allows for sound waves to travel the length of cochlea, which is between the oval and round windows.

VESTIBULE. The central part of the inner ear, housing the two otolith organs: the UTRICLE and the SACCULE.

- OTOLITH. Ear crystal.
- UTRICLE. An organ of the vestibular system, lying in the horizontal plane. A layer of otoliths suspended in gel atop a sea of hair cells. Senses tilts and linear accelerations (speeding up in a straight line), especially forward/back and side to side.
- SACCULE. Similar to the UTRICLE, but vertically oriented. Senses gravity and linear accelerations, especially up/down.

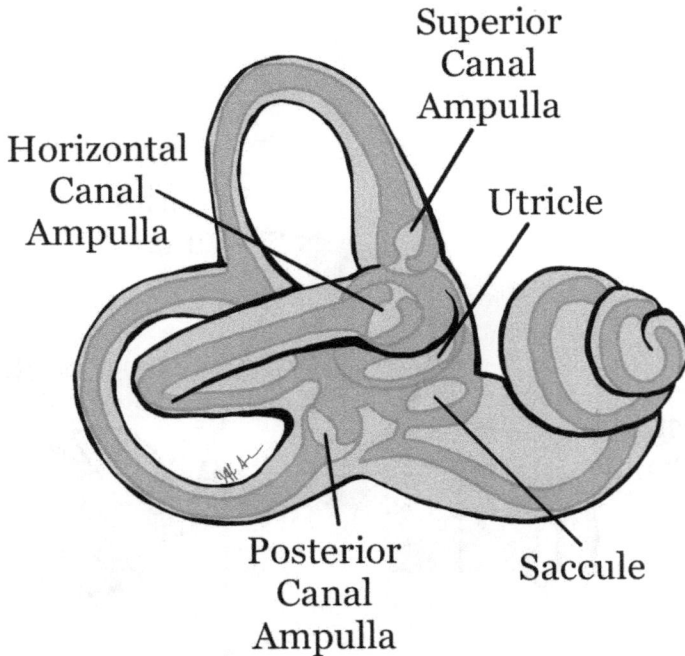

FIGURE 15.4 Details of the membranous labyrinth.

THE OTOLITH ORGANS

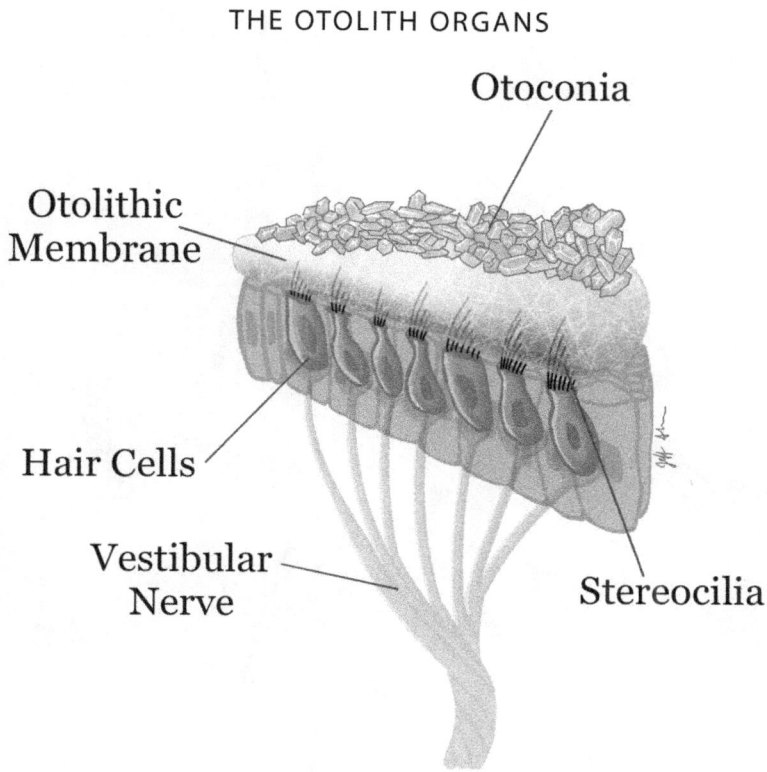

Otoconia

Otolithic
Membrane

Hair Cells

Vestibular
Nerve

Stereocilia

FIGURE 15.5 The otolith organs.

THE SEMICIRCULAR CANALS

Three canals, oriented at right angles, who sense angular acceleration (turning the head). All form a circular tube and have hair cells arranged within a platform called the CRISTA, with the hairs protruding into a gelatinous sail called the CUPULA. The AMPULLA is the widened part of the bony canal where the hair cells and cupula are located.

SUPERIOR CANAL. Also called the anterior canal, it senses diagonal downward head movements.

POSTERIOR CANAL. The lowermost canal, it senses diagonal upward head movements.

HORIZONTAL CANAL. Also called the lateral canal, it senses side-to-side head movements.

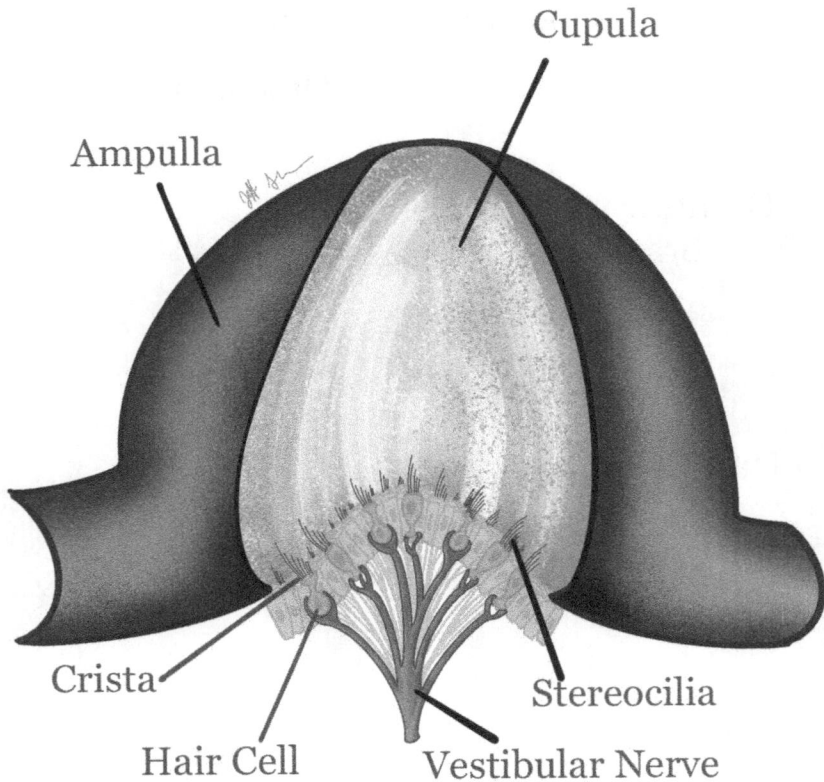

FIGURE 15.6 The ampulla of the semicircular canals.

THE VESTIBULAR SYSTEM

The parts of the inner ear (utricle, saccule, and three semicircular canals), the vestibular nerve, and the parts of the brain that aid in sensing movement.

VESTIBULO-OCULAR REFLEX. A reflex causing eyes to move in the opposite direction of head movements, to keep vision steady during motion.

VESTIBULOSPINAL REFLEX. A reflex controlling the body's muscles while moving, enabling balance and preventing falls.

NYSTAGMUS. A rhythmic twitching of the eyes. Vestibular nystagmus occurs when faulty vestibular signals are sent to the brain, resulting in a back and forth between a slow phase (the errant vestibular signal) and a fast phase (a resetting mechanism). Nystagmus is named for the fast phase, so a right-beating nystagmus means that the fast phase is to the right, and the slow phase is to the left.

GAIN. The ratio of head movement to compensatory eye movement. If the head moves to the right five degrees, and the eyes move to the left five degrees, then

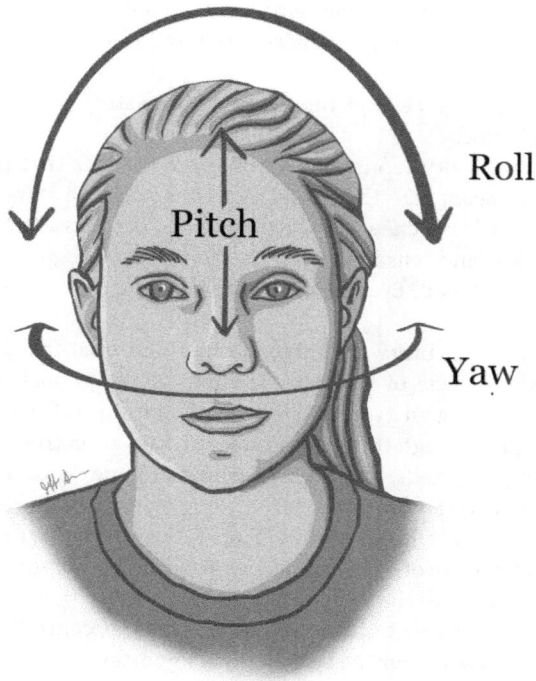

FIGURE 15.7 Yaw, pitch, and roll.

the gain is one. If eyes move less than the head, then the gain is less than one, and if they move more than the head, then the gain is greater than one.

VERTIGO. A false sensation of movement.

HEAD ROTATION. There are three planes of head rotation, similar to an airplane.

- YAW. Side-to-side head movement, like nodding "no."
- PITCH. Up and down head movement, like nodding "yes."
- ROLL. Tilting the head to the right or left, like trying to touch the shoulder with the ear.

THE CRANIAL NERVES

Twelve paired nerves, arising from the three parts of the brainstem, providing function to different parts of the head.

i. OLFACTORY NERVE. Smell.
ii. OPTIC NERVE. Sight.

iii. OCULOMOTOR NERVE. Eye movement (innervated four of six eye muscles, including the medial rectus, the superior and inferior rectus, and the inferior oblique).

iv. TROCHLEAR NERVE. Eye movement, innervating a single eye muscle, the superior oblique.

v. TRIGEMINAL NERVE. Supplies sensation to the face through three branches: V1—the forehead, V2—the cheeks, V3—the jaw. The third division also controls the muscles of chewing (the temporalis, the masseter, and the two pterygoid muscles), and sensation from the front of the tongue.

vi. ABDUCENS NERVE. Eye movement, controlling the final eye muscle, the lateral rectus.

vii. FACIAL NERVE. Every ear surgeon's first and final thought. The nerve that controls the muscle of the face—smiling, blinking, puckering, gnarling, and quizzically raising an eyebrow all depend on it. In addition, taste for the front of the tongue through the chorda tympani, salivation from the submandibular and sublingual glands, and tears (lacrimation) and nasal mucous through the greater superficial petrosal nerve.

viii. VESTIBULOCOCHLEAR NERVE. Hopefully by now this nerve needs no introduction! The nerve of hearing and balance, connecting the cochlea and the five vestibular organs to the brain.

ix. GLOSSOPHARYNGEAL NERVE. An eclectic nerve, controlling one muscle (styloglossus), sensation for the back of the tongue, taste for the back of the tongue, salivation for the parotid gland, and regulation of breathing and blood pressure.

x. VAGUS NERVE. From the Latin "The Wanderer." the Vagus nerve takes a circuitous course through the human body. In the head and neck region, it controls a few muscles (e.g., levator veli palatini), the voice box, and skin sensation around the ear canal (explaining the curious phenomenon whereby people cough when their ears get cleaned), throat sensation. Elsewhere in the body, it's critical for autonomic functions, deciding between sympathetic (fight or flight) and parasympathetic (rest and digest) modes. When the vagus is activated, parasympathetics take over, slowing heart rate, calming breathing, and preparing the stomach and intestines for an incoming meal.

xi. SPINAL ACCESSORY NERVE. Innervating two muscles that turn and tilt the neck: the sternocleidomastoid and the trapezius.

xii. GLOSSOPHARYNGEAL NERVE. Controls all the muscles inside the tongue.

TYPES OF EYE MOVEMENTS

SMOOTH PURSUIT. Eyes track a moving target, like a car passing by.

SACCADES. Eyes jump to a new target, like the source of a crashing sound.

OPTOKINETIC. Eyes move in tandem with movement of the entire visual world, like seeing a train pass by out of the window of another train.

VESTIBULAR. Eyes move reflexively to counteract head turns, resulting in no net movement of the eyes, keeping vision steady during movement.

BALANCE

The art of not falling, especially during movement. It's a multisensory achievement, with contributions from the following.

VISUAL. Sight.
VESTIBULAR. Inner ear accelerometers.
AUDITORY. Hearing.
SOMATOSENSORY. Somatosensory is a category heading that includes both proprioception and touch.

- PROPRIOCEPTION. The sense of knowing your own body position in space. Based on stretch sensors in muscles, it knows how bent each joint in the body is and is therefore able to calculate our pose at each moment in time.
- TOUCH. The ability to feel pressure against the skin.

VESTIBULAR DISEASES

BENIGN PAROXYSMAL POSITIONAL VERTIGO (BPPV). A very common cause of brief vertigo, lasting seconds, triggered by changing position, most commonly rolling over in bed. Caused by loose crystals (otoconia) relocating to the semicircular canals—usually the posterior canal. The crystals cause a massive cupular deflection with small head movements, tricking the brain into thinking that the whole body is rapidly cartwheeling, causing nystagmus and vertigo. See chapter 8.
VESTIBULAR MIGRAINE. A type of migraine that causes dizziness and vertigo, sometimes with a typical migraine headache, and sometimes without a migraine headache. This disease is underdiagnosed and underappreciated. At its worst, symptoms can last years and can be disabling. See chapter 9.
MÉNIÈRE'S DISEASE. This is a degenerative condition of the inner ear, causing hearing loss and vertigo. Typically, one ear is affected. Pressure in the ear and tinnitus are also common. The cause is unknown. See chapter 9.
PERSISTENT POSTURAL PERCEPTUAL DIZZINESS (PPPD). PPPD is thought to be a vestibular processing disorder. After an initial vestibular problem, the brain tries to adapt. However, in some people, the adaptation can backfire, perpetuating symptoms for months or years.
SUPERIOR CANAL DEHISCENCE SYNDROME (SCDS). This is a rare disease that was only recently discovered. An opening in the bony shell of the inner ear wreaks havoc inside, causing bizarre symptoms, including hearing your eyes move, and imbalance with loud sounds. See chapter 11.

VESTIBULAR NEURITIS. The balance nerve normally sends millions of signals to the brain each second. With a presumed viral inflammation that grinds to a halt, tricking the brain into thinking that you are rapidly spinning.

VESTIBULAR SCHWANNOMA. A benign (noncancerous) brain tumor that typically grows from the vestibular nerve, typically causing dizziness and hearing loss on the affected side.

STROKE. The brain relies on a constant supply of sugar and oxygen delivered with blood. Arteries, the blood pipes of the body, can clog or burst, causing tissue death within minutes. Different parts of the brain have different jobs and different arteries, so the exact neurologic problem—like a paralyzed face or a numb leg—will depend on the affected brain region.

UNILATERAL VESTIBULAR LOSS. Losing the vestibular system in just one ear. This can be caused by some other diseases mentioned here, like vestibular neuritis or vestibular schwannoma. In most cases, with good physical therapy, people can compensate without a noticeable decline in function.

BILATERAL VESTIBULAR LOSS. See chapter 10. Losing the vestibular system in both ears. This is a worse problem than just losing function in one ear. Those affected have impaired balance and a high fall risk, especially in situations with low light and uneven terrain.

LABYRINTHITIS. Inflammation in the inner ear. In severe cases, where bacteria gain entry into the cloistered labyrinth, hearing and balance can be permanently lost.

MAL DE DÉBARQUEMENT SYNDROME (MDDS). French for the sickness of disembarking. Everyone adapts to the roll of ocean waves, but some cannot seem to readapt to dry land. Instead, a constant rocking is felt for months or years, as though they were still at sea.

Notes

INTRODUCTION: A SENSE OF WONDER

1. Martha W. Bagnall and David Schoppik, "Development of Vestibular Behaviors in Zebrafish," *Current Opinion in Neurobiology* 53 (December 2018): 83–89; B. B. Riley and S. J. Moorman, "Development of Utricular Otoliths, But Not Saccular Otoliths, Is Necessary for Vestibular Function and Survival in Zebrafish," *Journal of Neurobiology* 43, no. 4 (2000): 329–37.

1. AN AURAL HISTORY

1. Jan Hiller et al., "Assessing Inner Ear Volumetric Measurements by Using Three-Dimensional Reconstruction Imaging of High-Resolution Cone-Beam Computed Tomography," *SN Comprehensive Clinical Medicine* 2, no. 11 (2020): 2178–84.
2. Nicholas Culpeper, *The English Physician* (1708; University of Alabama Press, 2007).
3. Adam Politzer, *History of Otology: From Earliest Times to the Middle of the Nineteenth Century* (Columella Press, 1981).
4. Robert Baloh, "Prosper Ménière and His Disease," *Archives of Neurology* 58, no. 7 (2001): 1151–56.
5. V. Henn and L. R. Young, "Ernst Mach on the Vestibular Organ 100 Years Ago," *ORL: Journal for Oto-Rhino-Laryngology and Its Related Specialties* 37, no. 3 (1975): 138–48; Gerald Wiest and Robert W. Baloh, "The Pioneering Work of Josef Breuer on the Vestibular System," *Archives of Neurology* 59, no. 10 (October 2002): 1647–53.
6. Henn and Young, "Ernst Mach on the Vestibular Organ 100 Years Ago."
7. Wiest and Baloh, "The Pioneering Work of Josef Breuer on the Vestibular System."

225

2. HOW CHEETAHS PROSPER (EVOLUTION)

1. Jorge Luis Borges, "The Library of Babel" (Penguin Classics, 2023).
2 Charles Darwin and Leonard Kebler, *On the Origin of Species by Means of Natural Selection, or, The Preservation of Favoured Races in the Struggle for Life* (J. Murray, 1859). https://www.loc.gov/item/06017473/.
3. Bernd Fritzsch and Hans Straka, "Evolution of Vertebrate Mechanosensory Hair Cells and Inner Ears: Toward Identifying Stimuli That Select Mutation Driven Altered Morphologies," *Journal of Comparative Physiology A: Neuroethology, Sensory, Neural, and Behavioral Physiology* 200, no. 1 (January 2014): 5–18.
4. Fritsch and Straka, "Evolution of Vertebrate Mechanosensory Hair Cells and Inner Ears."
5. Christopher S. von Bartheld and Francesco Giannessi, "The Paratympanic Organ: A Barometer and Altimeter in the Middle Ear of Birds?," *Journal of Experimental Zoology Part B: Molecular and Developmental Evolution* 316, no. 6 (September 2011): 402–8.
6. Barany Society 2022 Keynote Lecture, "The Vestibular Ocular Reflex: An Evolutionarily Conserved Vertebrate Behavior."
7. Lawrence M. Witmer and Ryan C. Ridgely, "New Insights into the Brain, Braincase, and Ear Region of Tyrannosaurs (Dinosauria, Theropoda), with Implications for Sensory Organization and Behavior," *Anatomical Record* 292, no. 9 (September 2009): 1266–96.
8. Le-Qing Wu and J. David Dickman, "Neural Correlates of a Magnetic Sense," *Science* 336, no. 6084 (May 2012): 1054–57.
9. G. A. Manley and C. Köppl, "Phylogenetic Development of the Cochlea and Its Innervation," *Current Opinion in Neurobiology* 8, no. 4 (August 1998): 468–74.
10. Camille Grohé, Beatrice Lee, and John J. Flynn, "Recent Inner Ear Specialization for High-Speed Hunting in Cheetahs," *Scientific Reports* 8, no. 1 (February 2018): 2301.
11. Fred Spoor et al., "The Primate Semicircular Canal System and Locomotion," *Proceedings of the National Academy of Sciences of the United States of America* 104, no. 26 (June 2007): 10808–12.
12. F. Spoor, B. Wood, and F. Zonneveld, "Implications of Early Hominid Labyrinthine Morphology for Evolution of Human Bipedal Locomotion," *Nature* 369, no. 6482 (June 1994): 645–48.
13. Richard L. Essner Jr., et al., "Semicircular Canal Size Constrains Vestibular Function in Miniaturized Frogs," *Science Advances* 8, no. 24 (June 2022): eabn1104.
14. Essner Jr., et al., "Semicircular Canal Size," supplementary materials, https://www.science.org/doi/10.1126/sciadv.abn1104#supplementary-materials.
15. John A. Bender and Mark A. Frye, "Invertebrate Solutions for Sensing Gravity," *Current Biology: CB* 19, no. 5 (March 2009): R186–90.

3. A HEAD FULL OF HAIR CELLS

1. Saumil N. Merchant and Joseph B. Nadol, eds., *Schuknect's Pathology of the Ear*, 3rd ed. (pmph usa, 2010), chap. 3.
2. Aldous Huxley, *Point Counter Point* (The Modern Library, 1928).
3. Jay M. Goldberg, *The Vestibular System: A Sixth Sense* (Oxford University Press, 2012).
4. U. Rosenhall, "Mapping of the Cristae Ampullares in Man," *Annals of Otology, Rhinology, and Laryngology* 81, no. 6 (December 1972): 882–89.
5. Lawrence R. Lustig, et al., *Clinical Neurotology: Diagnosing and Managing Disorders of Hearing, Balance and the Facial Nerve* (CRC Press, 2002), chap. 3.
6. C. M. Oman, E. N. Marcus, and I. S. Curthoys, "The Influence of Semicircular Canal Morphology on Endolymph Flow Dynamics. An Anatomically Descriptive Mathematical Model," *Acta Oto-Laryngologica* 103, nos. 1–2 (January–February 1987): 1–13.
7. Jérome Carriot, et al., "Statistics of the Vestibular Input Experienced during Natural Self-Motion: Implications for Neural Processing," *Journal of Neuroscience: The Official Journal of the Society for Neuroscience* 34, no. 24 (June 2014): 8347–57.
8. Rosenhall, "Mapping of the Cristae Ampullares in Man."
9. Stefano Ramat and David S. Zee, "Ocular Motor Responses to Abrupt Interaural Head Translation in Normal Humans," *Journal of Neurophysiology* 90, no. 2 (August 2003): 887–902.
10. D. E. Angelaki, et al., "Computation of Inertial Motion: Neural Strategies to Resolve Ambiguous Otolith Information," *Journal of Neuroscience: The Official Journal of the Society for Neuroscience* 19, no. 1 (January 1999): 316–27.
11. Christopher Zalewski, *Rotational Vestibular Assessment* (Plural Publishing, 2017), 108.

4. THE EYES HAVE IT: VESTIBULO-OCULAR REFLEX

1. Marko Huterer and Kathleen E. Cullen, "Vestibuloocular Reflex Dynamics During High-Frequency and High-Acceleration Rotations of the Head on Body in Rhesus Monkey," *Journal of Neurophysiology* 88, no. 1 (July 2002): 13–28.
2. G. M. Halmagyi and I. S. Curthoys, "A Clinical Sign of Canal Paresis," *Archives of Neurology* 45, no. 7 (July 1988): 737–39.
3. J. G. Colebatch, G. M. Halmagyi, and N. F. Skuse, "Myogenic Potentials Generated by a Click-Evoked Vestibulocollic Reflex," *Journal of Neurology, Neurosurgery, and Psychiatry* 57, no. 2 (February 1994): 190–97.

5. THE BRAINS BEHIND THE OPERATION

1. L. Velázquez-Villaseñor, et al., "Temporal Bone Studies of the Human Peripheral Vestibular System. Normative Scarpa's Ganglion Cell Data," *Annals of Otology, Rhinology, and Laryngology*, Supplement 181 (May 2000): 14–19.

2. Jorge C. Kattah, et al., "HINTS to Diagnose Stroke in the Acute Vestibular Syndrome: Three-Step Bedside Oculomotor Examination More Sensitive than Early MRI Diffusion-Weighted Imaging," *Stroke: A Journal of Cerebral Circulation* 40, no. 11 (November 2009): 3504–10.

3. H. Rambold, et al., "Partial Ablations of the Flocculus and Ventral Paraflocculus in Monkeys Cause Linked Deficits in Smooth Pursuit Eye Movements and Adaptive Modification of the VOR," *Journal of Neurophysiology* 87, no. 2 (February 2002): 912–24.

4. Pierre Sachse, et al., "'The World Is Upside Down'—The Innsbruck Goggle Experiments of Theodor Erismann (1883–1961) and Ivo Kohler (1915–1985)," *Cortex: A Journal Devoted to the Study of the Nervous System and Behavior* 92 (July 1, 2017): 222–32.

5. Mohammad Aleisa, Anthony G. Zeitouni, and Kathleen E. Cullen, "Vestibular Compensation After Unilateral Labyrinthectomy: Normal Versus Cerebellar Dysfunctional Mice," *Journal of Otolaryngology* 36, no. 6 (December 2007): 315–21.

6. M. Fetter and J. Dichgans, "Adaptive Mechanisms of VOR Compensation after Unilateral Peripheral Vestibular Lesions in Humans," *Journal of Vestibular Research: Equilibrium & Orientation* 1, no. 1 (1990): 9–22.

7. Gregory C. DeAngelis and Dora E. Angelaki, "Visual–Vestibular Integration for Self-Motion Perception," in *The Neural Bases of Multisensory Processes*, ed. Micah M. Murray and Mark T. Wallace (CRC Press/Taylor & Francis, 2012).

8. Thomas Brandt and Marianne Dieterich, "The Dizzy Patient: Don't Forget Disorders of the Central Vestibular System," *Nature Reviews. Neurology* 13, no. 6 (June 2017): 352–62.

9. Xun-Bei Shi et al., "Whole-Brain Monosynaptic Outputs and Presynaptic Inputs of GABAergic Neurons in the Vestibular Nuclei Complex of Mice," *Frontiers in Neuroscience* 16 (August 2022): 982596.

6. DON'T FALL FOR IT: VESTIBULOSPINAL REFLEXES

1. Rudolf Magnus, "Wie sich die fallende Katze in der Luft umdreht, Archs Néerl," *Physiol* (1922).

2. Yuri Agrawal, et al., "Head Impulse Test Abnormalities and Influence on Gait Speed and Falls in Older Individuals," *Otology & Neurotology: Official Publication of the American Otological Society, American Neurotology Society [and] European Academy of Otology and Neurotology* 34, no. 9 (December 2013): 1729–35.

3. David A. Ganz and Nancy K. Latham, "Prevention of Falls in Community-Dwelling Older Adults," *New England Journal of Medicine* 382, no. 8 (2020): 734–43.

4. Alex Tumarkin, "The Otolithic Catastrophe: A New Syndrome," *British Medical Journal* 2, no. 3942 (July 1936): 175–77.

5. R. Greenwood and A. Hopkins, "Muscle Responses during Sudden Falls in Man," *Journal of Physiology* 254, no. 2 (January 1976): 507–18.

6. J. G. Colebatch, G. M. Halmagyi, and N. F. Skuse, "Myogenic Potentials Generated by a Click-Evoked Vestibulocollic Reflex," *Journal of Neurology, Neurosurgery, and Psychiatry* 57, no. 2 (February 1994): 190–97.

7. David E. Ehrlich and David Schoppik, "A Primal Role for the Vestibular Sense in the Development of Coordinated Locomotion," *eLife* 8 (October 2019), doi:10.7554/eLife.45839.

8. Kavelin Rumalla, Adham M. Karim, and Timothy E. Hullar, "The Effect of Hearing Aids on Postural Stability," *The Laryngoscope* 125, no. 3 (March 2015): 720–23.

7. A BALANCED MIND: MEMORY AND COGNITION

1. Paul F. Smith, Cynthia L. Darlington, and Yiwen Zhen, "The Effects of Complete Vestibular Deafferentation on Spatial Memory and the Hippocampus in the Rat: The Dunedin Experience," *Multisensory Research* 28, no. 5–6 (2015): 461–85.

2. John O'Keefe, *The Hippocampus as a Cognitive Map* (Oxford University Press, 1978).

3. Dora E. Angelaki, et al., "A Gravity-Based Three-Dimensional Compass in the Mouse Brain," *Nature Communications* 11, no. 1 (April 2020): 1855.

4. Noah A. Russell, et al., "Long-Term Effects of Permanent Vestibular Lesions on Hippocampal Spatial Firing," *Journal of Neuroscience: The Official Journal of the Society for Neuroscience* 23, no. 16 (July 2003): 6490–98.

5. Thomas Brandt, et al., "Vestibular Loss Causes Hippocampal Atrophy and Impaired Spatial Memory in Humans," *Brain: A Journal of Neurology* 128, no. 11 (November 2005): 2732–41.

6. E. A. Maguire, et al., "Navigation-Related Structural Change in the Hippocampi of Taxi Drivers," *Proceedings of the National Academy of Sciences of the United States of America* 97, no. 8 (April 2000): 4398–4403.

7. Christophe Lopez and Maya Elzière, "Out-of-Body Experience in Vestibular Disorders—A Prospective Study of 210 Patients with Dizziness," *Cortex: A Journal Devoted to the Study of the Nervous System and Behavior* 104 (July 2018): 193–206.

8. Olaf Blanke, et al., "Stimulating Illusory Own-Body Perceptions," *Nature* 419, no. 6904 (September 2002): 269–70.

9. Robin T. Bigelow, et al., "Vestibular Vertigo and Comorbid Cognitive and Psychiatric Impairment: The 2008 National Health Interview Survey," *Journal of Neurology, Neurosurgery, and Psychiatry* 87, no. 4 (April 2016): 367–72.
10. Author interview with Yuri Agrawal.
11. Author interview with Yuri Agrawal.
12. Adriano R. Lameira and Marcus Perlman, "Great Apes Reach Momentary Altered Mental States by Spinning," *Primates; Journal of Primatology* 64, no. 3 (May 2023): 319–23.
13. Yusuf O. Cakmak, et al., "A Possible Role of Prolonged Whirling Episodes on Structural Plasticity of the Cortical Networks and Altered Vertigo Perception: The Cortex of Sufi Whirling Dervishes," *Frontiers in Human Neuroscience* 11 (January 2017): 3.
14. Aurore A. Perrault, et al., "Whole-Night Continuous Rocking Entrains Spontaneous Neural Oscillations with Benefits for Sleep and Memory," *Current Biology: CB* 29, no. 3 (February 2019): 402–11.e3.
15. Habib G. Rizk, et al., "Cross-Sectional Analysis of Cognitive Dysfunction in Patients With Vestibular Disorders," *Ear and Hearing* 41, no. 4 (2020): 1020–27.
16. Barbara Tversky, *Mind in Motion: How Action Shapes Thought* (Basic Books, 2019).

8. LIKE A ROLLING STONE: BENIGN PAROXYSMAL POSITIONAL VERTIGO (BPPV)

1. J. S. Oghalai, et al., "Unrecognized Benign Paroxysmal Positional Vertigo in Elderly Patients," *Otolaryngology—Head and Neck Surgery: Official Journal of American Academy of Otolaryngology—Head and Neck Surgery* 122, no. 5 (May 2000): 630–34.
2. Fernando Freitas Ganança et al., "Elderly Falls Associated with Benign Paroxysmal Positional Vertigo," *Brazilian Journal of Otorhinolaryngology* 76, nos. 1 (January–February 2010): 113–20.
3. Robert W. Baloh, *Vertigo: Five Physician Scientists and the Quest for a Cure* (Oxford University Press, 2016).
4. M. R. Dix and C. S. Hallpike, "The Pathology, Symptomatology and Diagnosis of Certain Common Disorders of the Vestibular System," *Annals of Otology, Rhinology, and Laryngology* 61, no. 4 (December 1952): 987–1016.
5. H. F. Schuknecht, "Cupulolithiasis," *Archives of Otolaryngology* 90, no. 6 (December 1969): 765–78.
6. "Epley Maneuver for Vertigo was Invented by Oregon Doctor," OregonLive.com/*The Oregonian*, https://www.oregonlive.com/health/2019/10/eply-maneuver-for-vertigo-was-invented-by-oregon-doctor.html.
7. J. M. Epley, "The Canalith Repositioning Procedure: For Treatment of Benign Paroxysmal Positional Vertigo," *Otolaryngology—Head and Neck Surgery:*

Official Journal of American Academy of Otolaryngology—Head and Neck Surgery 107, no. 3 (September 1992): 399–404.

8. L. S. Parnes and J. A. McClure, "Posterior Semicircular Canal Occlusion for Intractable Benign Paroxysmal Positional Vertigo," *Annals of Otology, Rhinology, and Laryngology* 99, no. 5 (May 1990): 330–34.

9. Wee Tin K. Kao, Lorne S. Parnes, and Richard A. Chole, "Otoconia and Otolithic Membrane Fragments within the Posterior Semicircular Canal in Benign Paroxysmal Positional Vertigo," *The Laryngoscope* 127, no. 3 (March 2017): 709–14.

10. Hui Xu, et al., "Evaluation of the Utricular and Saccular Function Using oVEMPs and cVEMPs in BPPV Patients," *Journal of Otolaryngology—Head & Neck Surgery = Le Journal D'oto-Rhino-Laryngologie et de Chirurgie Cervico-Faciale* 45 (February 2016): 12.

9. UNSOLVED MYSTERIES: VESTIBULAR MIGRAINE AND MÉNIÈRE'S DISEASE

1. Thi A. Preysner, et al., "Vestibular Migraine: Cognitive Dysfunction, Mobility, Falls," *Otology & Neurotology: Official Publication of the American Otological Society, American Neurotology Society [and] European Academy of Otology and Neurotology* 43, no. 10 (December 2022): 1216.

2. J. M. S. Pearce, "The Neurology of Aretaeus: Radix Pedis Neurologia," *European Neurology* 70, nos. 1–2 (July 2013): 106–12.

3. Ariel Winnick, et al., "Errors of Upright Perception in Patients With Vestibular Migraine," *Frontiers in Neurology* 9 (October 2018): 892.

4. Z. Vass, et al., "Co-Localization of the Vanilloid Capsaicin Receptor and Substance P in Sensory Nerve Fibers Innervating Cochlear and Vertebro-Basilar Arteries," *Neuroscience* 124, no. 4 (2004): 919–27.

5. Sherri M. Jones, et al., "Loss of α-Calcitonin Gene-Related Peptide (αCGRP) Reduces Otolith Activation Timing Dynamics and Impairs Balance," *Frontiers in Molecular Neuroscience* 11 (August 2018): 289.

6. ClinicalTrials.gov, NCT02447991.

7. Miguel Maldonado Fernandez et al., "Pharmacological Agents for the Prevention of Vestibular Migraine," *Cochrane Database of Systematic Reviews* 6, no. 6 (2015), https://www.researchgate.net/profile/Michael_Strupp/publication /278969320_Pharmacological_agents_for_the_prevention_of_vestibular _migraine/links/5c33684d92851c22a3624f3c/Pharmacological-agents-for -the-prevention-of-vestibular-migraine.pdf.

8. Otmar Bayer, et al., "Results and Lessons Learnt from a Randomized Controlled Trial: Prophylactic Treatment of Vestibular Migraine with Metoprolol (PROVEMIG)," *Trials* 20, no. 1 (December 2019): 813.

9. "The Guest House," poem by Rumi.

10. Eric J. Formeister et al., "Mindfulness-Based Stress Reduction for the Treatment of Vestibular Migraine: A Prospective Pilot Study," *Cureus* 17, no. 2 (2025),

https://www.cureus.com/articles/284205-mindfulness-based-stress-reduction-for-the-treatment-of-vestibular-migraine-a-prospective-pilot-study?score_article=true#!/metrics.

11. UCFS Balance and Falls Center, www.ohns.ucsf.edu/balance-falls.

12. C. Eduardo Corrales and Albert Mudry, "History of the Endolymphatic Sac: From Anatomy to Surgery," *Otology & Neurotology: Official Publication of the American Otological Society, American Neurotology Society [and] European Academy of Otology and Neurotology* 38, no. 1 (January 2017): 152–56.

13. Alex Tumarkin, "The Otolithic Catastrophe: A New Syndrome," *British Medical Journal* 2, no. 3942 (July 1936).

14. Vincent van Gogh: The Letters, vangoghletters.org.

15. "To Theo van Gogh. Saint-Rémy-de-Provence, on or about Thursday, 23 May 1889," Vincent van Gogh: The Letters, https://vangoghletters.org/vg/letters/let776/letter.html.

10. SENSELESS: BILATERAL VESTIBULAR LOSS

1. M. E. Dandy, "The Surgical Treatment of Meniere's Disease," *Surgery, Gynecology & Obstetrics* 72 (1941): 421–25.

2. J. Crawford, "Living Without A Balancing Mechanism," *British Journal of Ophthalmology* 48, no. 7 (July 1964): 357–60.

3. Crawford, "Living Without A Balancing Mechanism," 357–60.

4. J. F. Golding, "Motion Sickness," *Handbook of Clinical Neurology* 137 (2016): 371–90.

5. B. S. Cheung, I. P. Howard, and K. E. Money, "Visually-Induced Sickness in Normal and Bilaterally Labyrinthine-Defective Subjects," *Aviation, Space, and Environmental Medicine* 62, no. 6 (June 1991): 527–31.

6. K. E. Money and B. S. Cheung, "Another Function of the Inner Ear: Facilitation of the Emetic Response to Poisons," *Aviation, Space, and Environmental Medicine* 54, no. 3 (March 1983): 208–11.

7. Col. Chris Hadfield, *An Astronaut's Guide to Life on Earth: What Going to Space Taught Me about Ingenuity, Determination, and Being Prepared for Anything* (Pan Books, 2013).

8. Bryan Ripple, "NAMRU-D Releases the Kraken," Wright-Patterson Air Force Base, June 20, 2016, https://www.wpafb.af.mil/News/Article-Display/Article/818426/namru-d-releases-the-kraken/.

9. Florence Lucieer, et al., "Bilateral Vestibular Hypofunction: Insights in Etiologies, Clinical Subtypes, and Diagnostics," *Frontiers In*, 2016, https://www.frontiersin.org/articles/10.3389/fneur.2016.00026/full.

11. A HOLE IN THE HEAD: SUPERIOR CANAL DEHISCENCE SYNDROME

1. L. B. Minor et al., "Sound- and/or Pressure-Induced Vertigo due to Bone Dehiscence of the Superior Semicircular Canal," *Archives of Otolaryngology: Head and Neck Surgery* 124, no. 3 (March 1998): 249–58.
2. C. V. Dalchow, et al., "Imaging of Ancient Egyptian Mummies' Temporal Bones with Digital Volume Tomography," *European Archives of Oto-Rhino-Laryngology: Official Journal of the European Federation of Oto-Rhino-Laryngological Societies: Affiliated with the German Society for Oto-Rhino-Laryngology—Head and Neck Surgery* 269, no. 10 (October 2012): 2277–84.
3. Saumil N. Merchant and John J. Rosowski, "Conductive Hearing Loss Caused by Third-Window Lesions of the Inner Ear," *Otology & Neurotology: Official Publication of the American Otological Society, American Neurotology Society [and] European Academy of Otology and Neurotology* 29, no. 3 (April 2008): 282–89.

12. RESTORING A SENSE OF BALANCE: THE VESTIBULAR IMPLANT

1. Bernard Cohen, Jun-Ichi Suzuki, and Morris B. Bender, "XVI Eye Movements from Semicircular Canal Nerve Stimulation in the Cat," *Annals of Otology, Rhinology, and Laryngology* 73, no. 1 (March 1964): 153–69.
2. W. Gong and D. M. Merfeld, "Prototype Neural Semicircular Canal Prosthesis Using Patterned Electrical Stimulation," *Annals of Biomedical Engineering* 28, no. 5 (May 2000): 572–81.
3. Jean-Philippe Guyot, et al., "Adaptation to Steady-State Electrical Stimulation of the Vestibular System in Humans," *Annals of Otology, Rhinology, and Laryngology* 120, no. 3 (March 2011): 143–49.
4. Conrad Wall III, Maria Izabel Kos, and Jean-Philippe Guyot, "Eye Movements in Response to Electric Stimulation of the Human Posterior Ampullary Nerve," *Annals of Otology, Rhinology, and Laryngology* 116, no. 5 (May 2007): 369–74.
5. Peter J. Boutros, et al., "Continuous Vestibular Implant Stimulation Partially Restores Eye-Stabilizing Reflexes," *JCI Insight* 4, no. 22 (November 2019), doi:10.1172/jci.insight.128397.
6. Margaret R. Chow, et al., "Posture, Gait, Quality of Life, and Hearing with a Vestibular Implant," *New England Journal of Medicine* 384, no. 6 (February 2021): 521–32.

13. VESTIBULAR ROGAINE: HAIR CELL REGENERATION

1. J. T. Corwin, "Postembryonic Production and Aging in Inner Ear Hair Cells in Sharks," *Journal of Comparative Neurology* 201, no. 4 (October 1981): 541–53.

2. J. T. Corwin, "Postembryonic Growth of the Macula Neglecta Auditory Detector in the Ray, Raja Clavata: Continual Increases in Hair Cell Number, Neural Convergence, and Physiological Sensitivity," *Journal of Comparative Neurology* 217, no. 3 (July 1983): 345–56.
3. J. P. Carey, A. F. Fuchs, and E. W. Rubel, "Hair Cell Regeneration and Recovery of the Vestibuloocular Reflex in the Avian Vestibular System," *Journal of Neurophysiology* 76, no. 5 (November 1996): 3301–12.
4. R. M. Hegstrom, et al., "Two Year's Experience with Periodic Hemodialysis in the Treatment of Chronic Uremia," *Transactions: American Society for Artificial Internal Organs* 8 (1962): 266–80.
5. "How Well Do Dogs and Other Animals Hear," Louisiana State University, https://www.lsu.edu/deafness/HearingRange.html.
6. Jun Lv, et al., "AAV1-hOTOF Gene Therapy for Autosomal Recessive Deafness 9: A Single-Arm Trial," *The Lancet* 403, no. 10441 (May 2024): 2317–25.

14. SPACING OUT

1. Col. Chris Hadfield, *An Astronaut's Guide to Life on Earth: What Going to Space Taught Me About Ingenuity, Determination, and Being Prepared for Anything* (Pan Books, 2013), 175.
2. William E. Thornton and Frederick Bonato, "Space Motion Sickness and Motion Sickness: Symptoms and Etiology," *Aviation, Space, and Environmental Medicine* 84, no. 7 (July 2013): 716–21.
3. Emma Hallgren, et al., "Dysfunctional Vestibular System Causes a Blood Pressure Drop in Astronauts Returning from Space," *Scientific Reports* 5 (December 16, 2015): 17627.
4. C. J. Bockisch and T. Haslwanter, "Three-Dimensional Eye Position during Static Roll and Pitch in Humans," *Vision Research* 41, no. 16 (July 2001): 2127–37.
5. Millard F. Reschke and Gilles Clément, "Vestibular and Sensorimotor Dysfunction during Space Flight," *Current Pathobiology Reports*, July 3, 2018, https://link.springer.com/article/10.1007/s40139-018-0173-y.
6. Charles Oman, "Spatial Orientation and Navigation in Microgravity," in *Spatial Processing in Navigation, Imagery and Perception*, ed. Fred Mast and Lutz Jäncke (Springer US, 2007), 209–47.
7. D. B. Spangenberg, et al., "Graviceptor Development in Jellyfish Ephyrae in Space and on Earth," *Advances in Space Research: The Official Journal of the Committee on Space Research* 14, no. 8 (1994): 317–25.
8. April E. Ronca, et al., "Orbital Spaceflight During Pregnancy Shapes Function of Mammalian Vestibular System," *Behavioral Neuroscience* 122, no. 1 (February 2008): 224–32.
9. Kerry D. Walton, et al., "The Effects of Microgravity on the Development of Surface Righting in Rats," *Journal of Physiology* 565, Pt 2 (June 2005): 593–608.

10. Hadfield, *An Astronaut's Guide to Life on Earth.*
11. C. M. Oman and M. J. Kulbaski, "Spaceflight Affects the 1-G Postrotatory Vestibulo-Ocular Reflex," *Advances in Oto-Rhino-Laryngology* 42 (1988): 5–8.

15. SPINNING OUT OF CONTROL: ADVICE FOR PATIENTS

1. UCSF Balance and Falls Center, Department of Otolaryngology—Head and Neck Surgery, University of California, San Francisco, www.ohns.ucsf.edu /balance-falls.
2. UCSF Balance and Falls Center, www.ohns.ucsf.edu/balance-falls.
3. UCSF Balance and Falls Center, www.ohns.ucsf.edu/balance-falls.
4. Vestibular Disorder Association (VEDA), www.vestibular.org.
5. Dizziness-and-Balance.com, https://www.dizziness-and-balance.com/.
6. House Ear Institute/House Institute Foundation, https://hifla.org.
7. Association of Migraine Disorders, https://www.migrainedisorders.org/.
8. Bárány Society, https://www.thebaranysociety.org/.

Bibliography

Agrawal, Yuri, Marcela Davalos-Bichara, Maria Geraldine Zuniga, and John P. Carey. "Head Impulse Test Abnormalities and Influence on Gait Speed and Falls in Older Individuals." *Otology & Neurotology: Official Publication of the American Otological Society, American Neurotology Society [and] European Academy of Otology and Neurotology* 34, no. 9 (December 2013): 1729–35.

Aleisa, Mohammad, Anthony G. Zeitouni, and Kathleen E. Cullen. "Vestibular Compensation after Unilateral Labyrinthectomy: Normal versus Cerebellar Dysfunctional Mice." *The Journal of Otolaryngology* 36, no. 6 (December 2007): 315–21.

Angelaki, D. E., M. Q. McHenry, J. D. Dickman, S. D. Newlands, and B. J. Hess. "Computation of Inertial Motion: Neural Strategies to Resolve Ambiguous Otolith Information." *Journal of Neuroscience: The Official Journal of the Society for Neuroscience* 19, no. 1 (January 1999): 316–27.

Angelaki, Dora E., Julia Ng, Amada M. Abrego, Henry X. Cham, Eftihia K. Asprodini, J. David Dickman, and Jean Laurens. "A Gravity-Based Three-Dimensional Compass in the Mouse Brain." *Nature Communications* 11, no. 1 (April 15, 2020): 1855.

Bagnall, Martha W., and David Schoppik. "Development of Vestibular Behaviors in Zebrafish." *Current Opinion in Neurobiology* 53 (December 2018): 83–89.

Baloh, Robert. "Prosper Ménière and His Disease." *Archives of Neurology* 58, no. 7 (July 1, 2001): 1151–56.

——. *Vertigo: Five Physician Scientists and the Quest for a Cure.* Oxford University Press, 2016.

Bartheld, Christopher S. von, and Francesco Giannessi. "The Paratympanic Organ: A Barometer and Altimeter in the Middle Ear of Birds?" *Journal of Experimental Zoology. Part B, Molecular and Developmental Evolution* 316, no. 6 (September 2011): 402–8.

Bayer, Otmar, Christine Adrion, Amani Al Tawil, Ulrich Mansmann, Michael Strupp, and PROVEMIG investigators. "Results and Lessons Learnt from a Randomized Controlled Trial: Prophylactic Treatment of Vestibular Migraine with Metoprolol (PROVEMIG)." *Trials* 20, no. 1 (December 2019): 813.

Bender, John A., and Mark A. Frye. "Invertebrate Solutions for Sensing Gravity." *Current Biology: CB* 19, no. 5 (March 2009): R186–90.

Bigelow, Robin T., Yevgeniy R. Semenov, Sascha du Lac, Howard J. Hoffman, and Yuri Agrawal. "Vestibular Vertigo and Comorbid Cognitive and Psychiatric Impairment: The 2008 National Health Interview Survey." *Journal of Neurology, Neurosurgery, and Psychiatry* 87, no. 4 (April 2016): 367–72.

Blanke, Olaf, Stéphanie Ortigue, Theodor Landis, and Margitta Seeck. "Stimulating Illusory Own-Body Perceptions." *Nature* 419, no. 6904 (September 2002): 269–70.

Bockisch, C. J., and T. Haslwanter. "Three-Dimensional Eye Position during Static Roll and Pitch in Humans." *Vision Research* 41, no. 16 (July 2001): 2127–37.

Borges, Jorge Luis. "The Library of Babel." Penguin Classics, 2023.

Boutros, Peter J., Desi P. Schoo, Mehdi Rahman, Nicolas S. Valentin, Margaret R. Chow, Andrianna I. Ayiotis, Brian J. Morris, et al. "Continuous Vestibular Implant Stimulation Partially Restores Eye-Stabilizing Reflexes." *JCI Insight* 4, no. 22 (November 2019). doi:10.1172/jci.insight.128397.

Brandt, Thomas, and Marianne Dieterich. "The Dizzy Patient: Don't Forget Disorders of the Central Vestibular System." *Nature Reviews: Neurology* 13, no. 6 (June 2017): 352–62.

Brandt, Thomas, Franz Schautzer, Derek A. Hamilton, Roland Brüning, Hans J. Markowitsch, Roger Kalla, Cynthia Darlington, Paul Smith, and Michael Strupp. "Vestibular Loss Causes Hippocampal Atrophy and Impaired Spatial Memory in Humans." *Brain: A Journal of Neurology* 128, Pt 11 (November 2005): 2732–41.

Cakmak, Yusuf O., Gazanfer Ekinci, Armin Heinecke, and Safiye Çavdar. "A Possible Role of Prolonged Whirling Episodes on Structural Plasticity of the Cortical Networks and Altered Vertigo Perception: The Cortex of Sufi Whirling Dervishes." *Frontiers in Human Neuroscience* 11 (January 2017): 3.

Carey, J. P., A. F. Fuchs, and E. W. Rubel. "Hair Cell Regeneration and Recovery of the Vestibuloocular Reflex in the Avian Vestibular System." *Journal of Neurophysiology* 76, no. 5 (November 1996): 3301–12.

Carriot, Jérome, Mohsen Jamali, Maurice J. Chacron, and Kathleen E. Cullen. "Statistics of the Vestibular Input Experienced during Natural Self-Motion: Implications for Neural Processing." *Journal of Neuroscience: The Official Journal of the Society for Neuroscience* 34, no. 24 (June 2014): 8347–57.

Cheung, B. S., I. P. Howard, and K. E. Money. "Visually-Induced Sickness in Normal and Bilaterally Labyrinthine-Defective Subjects." *Aviation, Space, and Environmental Medicine* 62, no. 6 (June 1991): 527–31.

Chow, Margaret R., Andrianna I. Ayiotis, Desi P. Schoo, Yoav Gimmon, Kelly E. Lane, Brian J. Morris, Mehdi A. Rahman, et al. "Posture, Gait, Quality of Life, and Hearing with a Vestibular Implant." *New England Journal of Medicine* 384, no. 6 (February 11, 2021): 521–32.

Cohen, Bernard, Jun-Ichi Suzuki, and Morris B. Bender. "XVI Eye Movements from Semicircular Canal Nerve Stimulation in the Cat." *Annals of Otology, Rhinology, and Laryngology* 73, no. 1 (March 1, 1964): 153–69.

Colebatch, J. G., G. M. Halmagyi, and N. F. Skuse. "Myogenic Potentials Generated by a Click-Evoked Vestibulocollic Reflex." *Journal of Neurology, Neurosurgery, and Psychiatry* 57, no. 2 (February 1994): 190–97.

Corrales, C. Eduardo, and Albert Mudry. "History of the Endolymphatic Sac: From Anatomy to Surgery." *Otology & Neurotology: Official Publication of the American Otological Society, American Neurotology Society [and] European Academy of Otology and Neurotology* 38, no. 1 (January 2017): 152–56.

Corwin, J. T. "Postembryonic Growth of the Macula Neglecta Auditory Detector in the Ray, Raja Clavata: Continual Increases in Hair Cell Number, Neural Convergence, and Physiological Sensitivity." *Journal of Comparative Neurology* 217, no. 3 (July 1, 1983): 345–56.

——. "Postembryonic Production and Aging in Inner Ear Hair Cells in Sharks." *The Journal of Comparative Neurology* 201, no. 4 (October 1, 1981): 541–53.

Crawford, J. "Living without a Balancing Mechanism." *British Journal of Ophthalmology* 48, no. 7 (July 1964): 357–60.

Culpeper, Nicholas. *The English Physician.* 1708; University of Alabama Press, 2007.

Dalchow, C. V., C. Schmidt, J. Harbort, R. Knecht, U. Grzyska, and A. Muenscher. "Imaging of Ancient Egyptian Mummies' Temporal Bones with Digital Volume Tomography." *European Archives of Oto-Rhino-Laryngology: Official Journal of the European Federation of Oto-Rhino-Laryngological Societies: Affiliated with the German Society for Oto-Rhino-Laryngology—Head and Neck Surgery* 269, no. 10 (October 2012): 2277–84.

Dandy, M. E. "The Surgical Treatment of Meniere's Disease." *Surgery, Gynecology & Obstetrics* 72 (1941): 421–25.

Darwin, Charles, and Leonard Kebler. *On the Origin of Species by Means of Natural Selection, or, The Preservation of Favoured Races in the Struggle for Life.* J. Murray, 1859. https://www.loc.gov/item/06017473/.

DeAngelis, Gregory C., and Dora E. Angelaki. "Visual–Vestibular Integration for Self-Motion Perception." In *The Neural Bases of Multisensory Processes,* ed. Micah M. Murray and Mark T. Wallace. CRC Press/Taylor & Francis, 2012.

Dix, M. R., and C. S. Hallpike. "The Pathology, Symptomatology and Diagnosis of Certain Common Disorders of the Vestibular System." *Annals of Otology, Rhinology, and Laryngology* 61, no. 4 (December 1952): 987–1016.

Ehrlich, David E., and David Schoppik. "A Primal Role for the Vestibular Sense in the Development of Coordinated Locomotion." *eLife* 8 (October 8, 2019). doi:10.7554/eLife.45839.

Epley, J. M. "The Canalith Repositioning Procedure: For Treatment of Benign Paroxysmal Positional Vertigo." *Otolaryngology—Head and Neck Surgery: Official Journal of American Academy of Otolaryngology—Head and Neck Surgery* 107, no. 3 (September 1992): 399–404.

Essner, Richard L., Jr, Rudá E. E. Pereira, David C. Blackburn, Amber L. Singh, Edward L. Stanley, Mauricio O. Moura, André E. Confetti, and Marcio R. Pie.

"Semicircular Canal Size Constrains Vestibular Function in Miniaturized Frogs." *Science Advances* 8, no. 24 (June 17, 2022): eabn1104.

Fetter, M., and J. Dichgans. "Adaptive Mechanisms of VOR Compensation after Unilateral Peripheral Vestibular Lesions in Humans." *Journal of Vestibular Research: Equilibrium & Orientation* 1, no. 1 (1990): 9–22.

Formeister, Eric. J., James Mitchell, Roseanne Krauter, Ricky Chae, Adam Gardi, Maxwell Hum, and Jeffrey D. Sharon. "Mindfulness-Based Stress Reduction for the Treatment of Vestibular Migraine: A Prospective Pilot Study." *Cureus* 17, no. 2 (2025). https://www.cureus.com/articles/284205-mindfulness-based -stress-reduction-for-the-treatment-of-vestibular-migraine-a-prospective -pilot-study?score_article=true#!/metrics.

Fritzsch, Bernd, and Hans Straka. "Evolution of Vertebrate Mechanosensory Hair Cells and Inner Ears: Toward Identifying Stimuli That Select Mutation Driven Altered Morphologies." *Journal of Comparative Physiology A: Neuroethology, Sensory, Neural, and Behavioral Physiology* 200, no. 1 (January 2014): 5–18.

Ganança, Fernando Freitas, Juliana Maria Gazzola, Cristina Freitas Ganança, Heloísa Helena Caovilla, Maurício Malavasi Ganança, and Oswaldo Laércio Mendonça Cruz. "Elderly Falls Associated with Benign Paroxysmal Positional Vertigo." *Brazilian Journal of Otorhinolaryngology* 76, nos. 1 (January–February 2010): 113–20.

Ganz, David A., and Nancy K. Latham. "Prevention of Falls in Community-Dwelling Older Adults." *New England Journal of Medicine* 382, no. 8 (2020): 734–43.

Goldberg, Jay M. *The Vestibular System: A Sixth Sense*. Oxford University Press, 2012.

Golding, J. F. "Motion Sickness." *Handbook of Clinical Neurology* 137 (2016): 371–90.

Gong, W., and D. M. Merfeld. "Prototype Neural Semicircular Canal Prosthesis Using Patterned Electrical Stimulation." *Annals of Biomedical Engineering* 28, no. 5 (May 2000): 572–81.

Greenwood, R., and A. Hopkins. "Muscle Responses during Sudden Falls in Man." *Journal of Physiology* 254, no. 2 (January 1976): 507–18.

Grohé, Camille, Beatrice Lee, and John J. Flynn. "Recent Inner Ear Specialization for High-Speed Hunting in Cheetahs." *Scientific Reports* 8, no. 1 (February 2018): 2301.

Guyot, Jean-Philippe, Alain Sigrist, Marco Pelizzone, and Maria Izabel Kos. "Adaptation to Steady-State Electrical Stimulation of the Vestibular System in Humans." *Annals of Otology, Rhinology, and Laryngology* 120, no. 3 (March 2011): 143–49.

Hadfield, Chris, Col. *An Astronaut's Guide to Life on Earth: What Going to Space Taught Me about Ingenuity, Determination, and Being Prepared for Anything*. Pan Books, 2013.

Hallgren, Emma, Pierre-François Migeotte, Ludmila Kornilova, Quentin Delière, Erik Fransen, Dmitrii Glukhikh, Steven T. Moore, et al. "Dysfunctional

Vestibular System Causes a Blood Pressure Drop in Astronauts Returning from Space." *Scientific Reports* 5 (December 2015): 17627.

Halmagyi, G. M., and I. S. Curthoys. "A Clinical Sign of Canal Paresis." *Archives of Neurology* 45, no. 7 (July 1988): 737–39.

Hegstrom, R. M., J. S. Murray, J. P. Pendras, J. M. Burnell, and B. H. Scribner. "Two Year's Experience with Periodic Hemodialysis in the Treatment of Chronic Uremia." *Transactions: American Society for Artificial Internal Organs* 8 (1962): 266–80.

Henn, V., and L. R. Young. "Ernst Mach on the Vestibular Organ 100 Years Ago." *ORL; Journal for Oto-Rhino-Laryngology and Its Related Specialties* 37, no. 3 (1975): 138–48.

Hiller, Jan, Nour-Eldin Abdelrehim Nour-Eldin, Tatjana Gruber-Rouh, Iris Burck, Marc Harth, Timo Stöver, Thomas Vogl, and Nagy Naguib Naeem Naguib. "Assessing Inner Ear Volumetric Measurements by Using Three-Dimensional Reconstruction Imaging of High-Resolution Cone-Beam Computed Tomography." *SN Comprehensive Clinical Medicine* 2, no. 11 (November 2020): 2178–84.

Huterer, Marko, and Kathleen E. Cullen. "Vestibuloocular Reflex Dynamics during High-Frequency and High-Acceleration Rotations of the Head on Body in Rhesus Monkey." *Journal of Neurophysiology* 88, no. 1 (July 2002): 13–28.

Huxley, Aldous. *Point Counter Point*. The Modern Library, 1928.

Jones, Sherri M., Sarath Vijayakumar, Samantha A. Dow, Joseph C. Holt, Paivi M. Jordan, and Anne E. Luebke. "Loss of α-Calcitonin Gene-Related Peptide (αCGRP) Reduces Otolith Activation Timing Dynamics and Impairs Balance." *Frontiers in Molecular Neuroscience* 11 (August 2018): 289.

Kao, Wee Tin K., Lorne S. Parnes, and Richard A. Chole. "Otoconia and Otolithic Membrane Fragments within the Posterior Semicircular Canal in Benign Paroxysmal Positional Vertigo." *The Laryngoscope* 127, no. 3 (March 2017): 709–14.

Kattah, Jorge C., Arun V. Talkad, David Z. Wang, Yu-Hsiang Hsieh, and David E. Newman-Toker. "HINTS to Diagnose Stroke in the Acute Vestibular Syndrome: Three-Step Bedside Oculomotor Examination More Sensitive than Early MRI Diffusion-Weighted Imaging." *Stroke: A Journal of Cerebral Circulation* 40, no. 11 (November 2009): 3504–10.

Lameira, Adriano R., and Marcus Perlman. "Great Apes Reach Momentary Altered Mental States by Spinning." *Primates: Journal of Primatology* 64, no. 3 (May 2023): 319–23.

Lopez, Christophe, and Maya Elzière. "Out-of-Body Experience in Vestibular Disorders—A Prospective Study of 210 Patients with Dizziness." *Cortex; a Journal Devoted to the Study of the Nervous System and Behavior* 104 (July 2018): 193–206.

Lucieer, F., P. Vonk, N. Guinand, and R. Stokroos. "Bilateral Vestibular Hypofunction: Insights in Etiologies, Clinical Subtypes, and Diagnostics." *Frontiers in Neurology* (2016). https://www.frontiersin.org/articles/10.3389/fneur.2016.00026/full.

Lustig, Lawrence R., John K. Niparko, Lloyd B. Minor, and John S. Zee, eds. *Clinical Neurotology: Diagnosing and Managing Disorders of Hearing, Balance and the Facial Nerve.* CRC Press, 2002.

Lv, Jun, Hui Wang, Xiaoting Cheng, Yuxin Chen, Daqi Wang, Longlong Zhang, Qi Cao, et al. "AAV1-hOTOF Gene Therapy for Autosomal Recessive Deafness 9: A Single-Arm Trial." *The Lancet* 403, no. 10441 (May 2024): 2317–25.

Magnus, Rudolf. "Wie sich die fallende Katze in der Luft umdreht, Archs Néerl." *Physiol* (1922).

Maguire, E. A., D. G. Gadian, I. S. Johnsrude, C. D. Good, J. Ashburner, R. S. Frackowiak, and C. D. Frith. "Navigation-Related Structural Change in the Hippocampi of Taxi Drivers." *Proceedings of the National Academy of Sciences of the United States of America* 97, no. 8 (April 2000): 4398–4403.

Maldonado Fernandez, Miguel, Jasminder S. Birdi, Greg J. Irving, Louisa Murdin, Ilkka Kivekäs, Michael Strupp. "Pharmacological Agents for the Prevention of Vestibular Migraine." *Cochrane Database of Systematic Reviews* 6, no. 6 (2015). https://www.researchgate.net/profile/Michael_Strupp/publication/278969320 _Pharmacological_agents_for_the_prevention_of_vestibular_migraine/links /5c33684d92851c22a3624f3c/Pharmacological-agents-for-the-prevention-of -vestibular-migraine.pdf.

Manley, G. A., and C. Köppl. "Phylogenetic Development of the Cochlea and Its Innervation." *Current Opinion in Neurobiology* 8, no. 4 (August 1998): 468–74.

Merchant, Saumil N., and John J. Rosowski. "Conductive Hearing Loss Caused by Third-Window Lesions of the Inner Ear." *Otology & Neurotology: Official Publication of the American Otological Society, American Neurotology Society [and] European Academy of Otology and Neurotology* 29, no. 3 (April 2008): 282–89.

Merchant, Saumil N., and Joseph B. Nadol, eds. *Schuknect's Pathology of the Ear,* 3rd ed. pmph usa, 2010.

Minor, L. B., D. Solomon, J. S. Zinreich, and D. S. Zee. "Sound- And/or Pressure-Induced Vertigo due to Bone Dehiscence of the Superior Semicircular Canal." *Archives of Otolaryngology—Head & Neck Surgery* 124, no. 3 (March 1998): 249–58.

Money, K. E., and B. S. Cheung. "Another Function of the Inner Ear: Facilitation of the Emetic Response to Poisons." *Aviation, Space, and Environmental Medicine* 54, no. 3 (March 1983): 208–11.

Oghalai, J. S., S. Manolidis, J. L. Barth, M. G. Stewart, and H. A. Jenkins. "Unrecognized Benign Paroxysmal Positional Vertigo in Elderly Patients." *Otolaryngology—Head and Neck Surgery: Official Journal of American Academy of Otolaryngology—Head and Neck Surgery* 122, no. 5 (May 2000): 630–34.

Oman, Charles. "Spatial Orientation and Navigation in Microgravity." In *Spatial Processing in Navigation, Imagery and Perception,* ed. Fred Mast and Lutz Jäncke, 209–47. Springer US, 2007.

Oman, C. M., and M. J. Kulbaski. "Spaceflight Affects the 1-G Postrotatory Vestibulo-Ocular Reflex." *Advances in Oto-Rhino-Laryngology* 42 (1988): 5–8.

Oman, C. M., E. N. Marcus, and I. S. Curthoys. "The Influence of Semicircular Canal Morphology on Endolymph Flow Dynamics. An Anatomically Descriptive Mathematical Model." *Acta Oto-Laryngologica* 103, nos. 1–2 (January–February 1987): 1–13.

Parnes, L. S., and J. A. McClure. "Posterior Semicircular Canal Occlusion for Intractable Benign Paroxysmal Positional Vertigo." *Annals of Otology, Rhinology, and Laryngology* 99, no. 5, Pt 1 (May 1990): 330–34.

Perrault, Aurore A., Abbas Khani, Charles Quairiaux, Konstantinos Kompotis, Paul Franken, Michel Muhlethaler, Sophie Schwartz, and Laurence Bayer. "Whole-Night Continuous Rocking Entrains Spontaneous Neural Oscillations with Benefits for Sleep and Memory." *Current Biology: CB* 29, no. 3 (February 2019): 402–11.e3.

Politzer, Adam. *History of Otology: From Earliest Times to the Middle of the Nineteenth Century*. Columella Press, 1981.

Preysner, Thi A., Adam Z. Gardi, Sarah Ahmad, and Jeffrey D. Sharon. "Vestibular Migraine: Cognitive Dysfunction, Mobility, Falls." *Otology & Neurotology: Official Publication of the American Otological Society, American Neurotology Society [and] European Academy of Otology and Neurotology* 43, no. 10 (December 2022): 1216.

Ramat, Stefano, and David S. Zee. "Ocular Motor Responses to Abrupt Interaural Head Translation in Normal Humans." *Journal of Neurophysiology* 90, no. 2 (August 2003): 887–902.

Rambold, H., A. Churchland, Y. Selig, L. Jasmin, and S. G. Lisberger. "Partial Ablations of the Flocculus and Ventral Paraflocculus in Monkeys Cause Linked Deficits in Smooth Pursuit Eye Movements and Adaptive Modification of the VOR." *Journal of Neurophysiology* 87, no. 2 (February 2002): 912–24.

Reschke, Millard F., and Gilles Clément. "Vestibular and Sensorimotor Dysfunction during Space Flight." *Current Pathobiology Reports* (July 3, 2018). https://link.springer.com/article/10.1007/s40139-018-0173-y.

Riley, B. B., and S. J. Moorman. "Development of Utricular Otoliths, but Not Saccular Otoliths, is Necessary for Vestibular Function and Survival in Zebrafish." *Journal of Neurobiology* 43, no. 4 (June 2000): 329–37.

Rizk, Habib G., Jeffrey D. Sharon, Joshua A. Lee, Cameron Thomas, Shaun A. Nguyen, and Ted A. Meyer. "Cross-Sectional Analysis of Cognitive Dysfunction in Patients With Vestibular Disorders." *Ear and Hearing* 41, no. 4 (2020): 1020–27.

Ronca, April E., Bernd Fritzsch, Laura L. Bruce, and Jeffrey R. Alberts. "Orbital Spaceflight during Pregnancy Shapes Function of Mammalian Vestibular System." *Behavioral Neuroscience* 122, no. 1 (February 2008): 224–32.

Rosenhall, U. "Mapping of the Cristae Ampullares in Man." *Annals of Otology, Rhinology, and Laryngology* 81, no. 6 (December 1972): 882–89.

Rumalla, Kavelin, Adham M. Karim, and Timothy E. Hullar. "The Effect of Hearing Aids on Postural Stability." *The Laryngoscope* 125, no. 3 (March 2015): 720–23.

Russell, Noah A., Arata Horii, Paul F. Smith, Cynthia L. Darlington, and David K. Bilkey. "Long-Term Effects of Permanent Vestibular Lesions on Hippocampal Spatial Firing." *Journal of Neuroscience: The Official Journal of the Society for Neuroscience* 23, no. 16 (July 2003): 6490–98.

Sachse, Pierre, Ursula Beermann, Markus Martini, Thomas Maran, Markus Domeier, and Marco R. Furtner. " 'The World Is Upside Down'—The Innsbruck Goggle Experiments of Theodor Erismann (1883–1961) and Ivo Kohler (1915–1985)." *Cortex: A Journal Devoted to the Study of the Nervous System and Behavior* 92 (July 2017): 222–32.

Schuknecht, H. F. "Cupulolithiasis." *Archives of Otolaryngology* 90, no. 6 (December 1969): 765–78.

Shi, Xun-Bei, Jing Wang, Fei-Tian Li, Yi-Bo Zhang, Wei-Min Qu, Chun-Fu Dai, and Zhi-Li Huang. "Whole-Brain Monosynaptic Outputs and Presynaptic Inputs of GABAergic Neurons in the Vestibular Nuclei Complex of Mice." *Frontiers in Neuroscience* 16 (August 26, 2022): 982596.

Smith, Paul F., Cynthia L. Darlington, and Yiwen Zhen. "The Effects of Complete Vestibular Deafferentation on Spatial Memory and the Hippocampus in the Rat: The Dunedin Experience." *Multisensory Research* 28, no. 5–6 (2015): 461–85.

Spangenberg, D. B., T. Jernigan, C. Philput, and B. Lowe. "Graviceptor Development in Jellyfish Ephyrae in Space and on Earth." *Advances in Space Research: The Official Journal of the Committee on Space Research* 14, no. 8 (1994): 317–25.

Spoor, Fred, Theodore Garland Jr, Gail Krovitz, Timothy M. Ryan, Mary T. Silcox, and Alan Walker. "The Primate Semicircular Canal System and Locomotion." *Proceedings of the National Academy of Sciences of the United States of America* 104, no. 26 (June 2007): 10808–12.

Spoor, F., B. Wood, and F. Zonneveld. "Implications of Early Hominid Labyrinthine Morphology for Evolution of Human Bipedal Locomotion." *Nature* 369, no. 6482 (June 1994): 645–48.

Thornton, William E., and Frederick Bonato. "Space Motion Sickness and Motion Sickness: Symptoms and Etiology." *Aviation, Space, and Environmental Medicine* 84, no. 7 (July 2013): 716–21.

Tumarkin, Alex "The Otolithic Catastrophe: A New Syndrome." *British Medical Journal* 2, no. 3942 (July 1936): 175–77.

Tversky, Barbara. *Mind in Motion: How Action Shapes Thought.* Basic Books, 2019.

Vass, Z., C. F. Dai, P. S. Steyger, G. Jancsó, D. R. Trune, and A. L. Nuttall. "Co-Localization of the Vanilloid Capsaicin Receptor and Substance P in Sensory Nerve Fibers Innervating Cochlear and Vertebro-Basilar Arteries." *Neuroscience* 124, no. 4 (2004): 919–27.

Velázquez-Villaseñor, L., S. N. Merchant, K. Tsuji, R. J. Glynn, C. Wall III, and S. D. Rauch. "Temporal Bone Studies of the Human Peripheral Vestibular System. Normative Scarpa's Ganglion Cell Data." *Annals of Otology, Rhinology, and Laryngology,* Supplement 181 (May 2000): 14–19.

Wall, Conrad, III, Maria Izabel Kos, and Jean-Philippe Guyot. "Eye Movements in Response to Electric Stimulation of the Human Posterior Ampullary Nerve." *Annals of Otology, Rhinology, and Laryngology* 116, no. 5 (May 2007): 369–74.

Walton, Kerry D., Shannon Harding, David Anschel, Ya'el Tobi Harris, and Rodolfo Llinás. "The Effects of Microgravity on the Development of Surface Righting in Rats." *Journal of Physiology* 565, Pt 2 (June 1, 2005): 593–608.

Wiest, Gerald, and Robert W. Baloh. "The Pioneering Work of Josef Breuer on the Vestibular System." *Archives of Neurology* 59, no. 10 (October 2002): 1647–53.

Winnick, Ariel, Shirin Sadeghpour, Jorge Otero-Millan, Tzu-Pu Chang, and Amir Kheradmand. "Errors of Upright Perception in Patients with Vestibular Migraine." *Frontiers in Neurology* 9 (October 30, 2018): 892.

Witmer, Lawrence M., and Ryan C. Ridgely. "New Insights into the Brain, Brain-case, and Ear Region of Tyrannosaurs (Dinosauria, Theropoda), with Implications for Sensory Organization and Behavior." *Anatomical Record* 292, no. 9 (September 2009): 1266–96.

Wu, Le-Qing, and J. David Dickman. "Neural Correlates of a Magnetic Sense." *Science* 336, no. 6084 (May 2012): 1054–57.

Xu, Hui, Fa-Ya Liang, Liang Chen, Xi-Cheng Song, Michael Chi Fai Tong, Jiun Fong Thong, Qing-Quan Zhang, and Yan Sun. "Evaluation of the Utricular and Saccular Function Using oVEMPs and cVEMPs in BPPV Patients." *Journal of Otolaryngology—Head & Neck Surgery = Le Journal D'oto-Rhino-Laryngologie et de Chirurgie Cervico-Faciale* 45 (February 2016): 12.

Zalewski, Christopher. *Rotational Vestibular Assessment.* Plural Publishing, 2017.

Index